rosewood

rosewood

endangered species conservation and the rise of global China

ANNAH LAKE ZHU

Harvard University Press
Cambridge, Massachusetts
London, England
2022

Printed in the United States of America

First printing

Library of Congress Cataloging-in-Publication Data

Names: Zhu, Annah Lake, 1985- author.
Title: Rosewood : endangered species conservation and the rise of global China / Annah Lake Zhu.
Description: Cambridge, Massachusetts : Harvard University Press, 2022. | Includes bibliographical references and index.
Identifiers: LCCN 2021048395 | ISBN 9780674260276 (hardcover)
Subjects: LCSH: Dalbergia—China. | Dalbergia—Conservation—China. | Sustainable forestry—China. | Rosewood furniture—China. | Rosewood (Wood)—China. | Forest conservation—China.
Classification: LCC SD397.D32 Z48 2022 | DDC 634.9/73—dc23/eng/20211122
LC record available at https://lccn.loc.gov/2021048395

. . . George

contents

abbreviations

CIRAD French Agricultural Research Centre for International Development
CITES Convention on International Trade in Endangered Species of Wild Fauna and Flora
COBA *communauté de base* (French name for communities managing forests)
EIA Environmental Investigation Agency
FDI foreign direct investment
GDP gross domestic product
ICDP integrated conservation and development project
IUCN International Union for Conservation of Nature
NGO nongovernmental organization
REDD reducing emissions from deforestation and forest degradation
RMB *rénmínbì*
SADC Southern African Development Community
UN United Nations
UNEP United Nations Environment Programme
UNESCO United Nations Educational, Scientific and Cultural Organization
UNODC United Nations Office on Drugs and Crime
US United States
USAID United States Agency for International Development
VOI *vondron'olona ifotony* (Malagasy name for communities managing forests)
WCS Wildlife Conservation Society (formerly New York Zoological Society)
WTO World Trade Organization
WWF World Wildlife Fund

rosewood

introduction

China's most valued commodity

One summer night in 2010, a group of thieves broke into the Chinese Pavilion at Drottningholm Palace, the royal residence of the king of Sweden. An alarm was triggered, and a police unit responded within six minutes. Yet it was already too late; the thieves had managed to slip into the night undetected, along with six priceless Chinese artworks. The theft was the first in a chain of museum heists across Europe, all of which targeted Chinese artifacts and antiques that had been taken out of China by Western powers in previous centuries. Initially, museum curators were perplexed. Why target such specific objects rather than other works of equal or greater value? And why such well-known items, which would be exceedingly difficult to resell on the black market or display privately?

From a Western perspective, these thefts made no sense. The market value of stolen Chinese artworks is severely limited in Western countries, and, as the Stockholm police noted, the notoriety of the pieces would compromise their resale potential.[1] Part of the challenge would be that in displaying these works, the buyer would appear scandalously complicit in the crime. But this logic holds true only for the Western collector. Within a different cultural-economic and geopolitical milieu that follows different legal boundaries, these so-called thefts appear otherwise. They in fact do not register as thefts at all but rather as repatriations—repatriations, that is, with substantial resale potential, given China's booming markets in historical cultural objects. Not only could these particular artworks be resold and displayed; they would be worth exponentially more than they were worth a decade prior.[2] Such rare classical art forms—"stolen" or otherwise—provide one of the most promising avenues for investment in contemporary China.

This dynamic is similar to that of endangered species bought and sold in China. Experiencing a cultural renaissance, culturally important endangered species now set record prices in Chinese markets. Pangolin scales can be worth more than silver, rhino horn worth more than diamonds. Rosewood (*hóng mù* in Mandarin), wood from a group of precious hardwoods that grow throughout the tropics, can be worth its weight in gold.[3] As with pangolin scales and rhino horn, rosewood's value is both cultural and speculative. Yet rosewood is trafficked far more than all other endangered species bought in China. Rosewood, in fact, is the world's most trafficked group of illicit wildlife by value.[4] From 2005 to 2014, the illicit trade in rosewood amounted to more than ivory, rhino horn, and big cats put together.[5] Nearly all rosewood logs cut globally are now sent to China, fueling a US$26 billion market for furniture created in the classical style of the Ming Dynasty. More than Louis Vuitton handbags or Rolex watches, China's nouveau riche are buying traditional rosewood furniture.

As with traditional artworks and cultural artifacts, endangered species historically used in China are experiencing a cultural renaissance that is interpreted very differently inside the country and outside of it. Whereas Western audiences denounce the use and consumption of endangered wildlife for either global health or environmental reasons, many Chinese consumers emphasize these species' cultural appeal and medicinal power. Demands to stop the wildlife trade originating from outside China not only face a vastly different legal landscape but also are reconceived through vastly different geopolitical and cultural assumptions within the country. The "thefts" at Drottningholm, though not unique, demonstrate this disconnect most clearly. "The line between commerce and crime," a journalist reporting on the heists observed, "is a really thin and a really blurry one."[6] Upon closer scrutiny, however, that line appears not so much thin and blurry as distinct and doubled—drawn completely differently in two different places.

The rhino horn wine goblet from Drottningholm—one of the six coveted objects stolen from the palace—exemplifies how an object may vacillate between licit and illicit, virtuous and scandalous, depending on the geopolitical and cultural position of the observer. As an artistic object on display in Sweden, the rhino horn goblet represents to the Chinese collector an illicit act unpunished, a reminder of an ugly colonial past that has not altogether passed. Repatriated to a private gallery in China, however, to the Western collector it now appears as contraband, evidence of a brazen crime left unprosecuted. Its material, the ancient remains of a now endangered species, elicits even more divergent responses. Rhino horn, like ivory, is one of many Chinese cultural goods made from endangered species. In China, these endangered resources—like works of ancient art—stand for a deep cultural heritage and a strong investment. In the West, they are more likely to symbolize the follies of unmitigated resource consumption.

This disconnect becomes more clear when considering China's contemporary demand for rosewood. Rosewood became a hot cultural commodity in China during the past two decades. The wood is used to make furniture designed in a classical style dating back to the Ming (1368–1644) and Qing (1644–1912) Dynasties. Buying rosewood furniture is not simply an act of what Westerners often call "conspicuous consumption." It is a revival of a cherished imperial history, a time when elaborate rosewood furnishings were commissioned by emperors and became one of the foremost symbols of social status in dynastic China (Figure i.1). The style of furniture was—and continues to be—deeply associated with the wood with which it is made. At peak production in late imperial China, the soul of the craftsman was thought to become immortalized in the polished wood, demonstrating both the skill of the worker and the refined taste of the consumer destined to pass on a prized family heirloom from one generation to the next. In keeping with tradition, nearly all rosewood pieces made today contain ornate carvings of iconic Chinese symbols and imagery thought to bring luck and good fortune to the owner. When cut and polished, rosewood materializes both personal prosperity and cultural excellence.

Similar to rosewood but less prominently featured in the household, ornate ivory decorations, rhino horn goblets, porcelain dishes, scrolls, and paintings all became hallmarks of the Chinese social elite of late imperial China. Modern consumers of endangered species may be buying into this cultural history, but current consumption patterns and the glory of China's cultural past are not connected seamlessly. Nearly fifty years after the fall of the Qing Dynasty, rosewood, along with all classical Chinese cultural elements, was sharply devalued. During China's mid-twentieth-century Cultural Revolution, the same rosewood and ivory heirlooms that adorned the imperial palaces were violently confiscated and vilified as symbols of bourgeois oppression. Antique paintings were mutilated and disfigured. Porcelain was broken and burned alongside bourgeois books and rosewood furniture. All became radically stigmatized and practically worthless.

Today, as China embraces capitalism and seeks to redefine its modernity with strong reference to its cultural past, rosewood furniture and other cultural goods are being bought back—or in some cases "stolen" back—by individuals who endured the hardships of the Cultural Revolution. Classically styled rosewood furnishings manufactured from endangered tropical hardwoods adorn the homes of China's new urban elite. If they can afford it, antique dishware and other

Figure i.1. Ming Dynasty rosewood furniture and decorations from the Imperial Palace, Beijing, December 2014. © Annah Lake Zhu.

distinctly Chinese artistic items adorn their homes as well. Initially cultural, the draw has become financial as well. Hearkening back to China's dynastic past, these ornate imperial objects now represent a ready buying opportunity for investors interested more in financial appreciation than in cultural esteem. Indeed, the market for Chinese traditional objects has been flooded by a deluge of global capital in search of new productive outlets. A small porcelain "chicken cup" looted from the imperial Summer Palace sold for US$36 million in 2014, only to be drunk from in front of a stunned crowd by the taxi-driver-turned-billionaire who bought it.[7] A vase initially put up at auction for US$800,000 ended up selling for US$69.5 million half an hour later.[8] Buying back cultural items stolen from foreign powers or seized by Communist Red Guards has become "high-street fashion" among China's elite.[9] The same might be said of endangered species consumption. Both draw on a tumultuous cultural history that Western conservationists and Western museum curators likely know little about.

decentering the West

This book's foundational observation is that twenty-first-century global modernity is witnessing a decentering of Western orthodoxy and a push to reclaim both cultural heritage and environmental sovereignty through diverse means, including art repatriation and endangered species protection. The initially perplexing museum thefts—which could be seen as a type of repatriation—exemplify this. The rightful place of Chinese antiques, the Elgin Marbles, artifacts from sub-Saharan Africa, and so forth are all hotly contested. Why, people around the world are beginning to ask much more forcefully than before, are these thefts sanctioned and proudly displayed, whereas the items appropriated by Axis powers during World War II, for example, are returned? Similar questions arise about endangered species, as Chapter 5 will delve into in much greater detail. Who has the right to define the legality of trade in endangered species across borders, and in which cultural and economic milieus will these definitions of legality hold weight? China and other countries across the globe are pushing back against international laws and norms that privilege Western approaches to endangered species conservation.[10] Other approaches, based on sustainable use and harvest, are finding their way onto the international agenda.[11]

The decentering of Western orthodoxy is one of the defining features of the contemporary world, whether in terms of art, governance, economy, or law. Economically, the rise of China and the decentering of the West have become quite clear. Culturally, however, things are much less understood. What does China's rise mean for the cultural hegemony of the West? How should we make meaning of the cultural history of a country so profoundly different from that of the West? The cultural questions surrounding China's rise are only beginning to be explored, not to mention the impact that these imminent cultural upheavals will have on endangered species conservation and, more broadly speaking, global environmentalism.[12] It is important to look at China's global rise not only as a threat to endangered species but also as a decentering of Western approaches to their conservation. In the face of this environmental decentering, further questions arise. How do Western approaches to endangered species conservation reproduce histories of colonial conquest, and how might these approaches, too, be "repatriated"?

The chapters that follow will demonstrate that the meaning of endangered species, what they do and don't help us to imagine, differs according to one's vantage point. Of all endangered species, rosewood exemplifies this most profoundly. From the tropical forests where it is logged to the Chinese timber markets where it is sold at exorbitant prices, the significance of rosewood transforms dramatically. In both cases, rosewood is markedly different from its image in the West as an iconic endangered species. Understanding these differences and what they mean for the future of endangered species conservation is the goal of this book.

The inspiration for this book began not in China but in Madagascar. Madagascar is a bastion of the China-bound illicit rosewood trade. From 2005 to 2014, more than half of the total volume of rosewood seized globally came from Madagascar, making this type of rosewood the most traded group of illicit species.[13] Living on the island as a Peace Corps volunteer in 2010 and the beginning of 2011, I witnessed the consequences of this vast illicit trade. I was stationed in the northeastern region of the island in a city that came to be known as "the heart of the rosewood trade" during a time locals called "the time of the rosewood." I saw firsthand how the trade completely transformed life in that region of Madagascar.

Any new houses that were constructed or new motorbikes that buzzed on the streets—and there were many—were said to be effects of rosewood. New metal roofs and solar panels out in the countryside were also attributed to the booming trade. In town, any buildings with walls above one's head were rumored to contain the endangered logs piled on the other side, and any houses recently built were alleged to have a cache of logs buried underneath (a built-in retirement plan, I was told). Mansions and sport utility vehicles, young men traveling in the backs of trucks, parties in the street, and the presence of the military (who from time to time were sent to the region to shut it all down) were all marked as the consequences of rosewood. The phenomenon of "hot money" (*vola mafana*)—spending one's earnings in trivial sprees or creative acts, such as pasting it to chameleons or boiling it up in a pot and eating it as soup—became a notorious feature of the social landscape. I was ceaselessly amazed by how this singular resource could transform a region so profoundly.

As much as Western aid dominates the conservation scene in Madagascar, Chinese demand for rosewood was so great that by 2010, Mandarin had become "the language of money." Rosewood became what locals called a "famous tree" (*kakazo malaza*): forbidden to cut yet worth a fortune in faraway places. None of this, I quickly found out, made much sense from the perspective of those living in the forests where, given the right price, the tree is indeed cut. Many have asked me, why all this interest in rosewood, why all this interest in the forest? What will the *chinois* (Chinese) do with the logs they are buying, and what will the *vazaha* (foreigners, typically White) do with the forest they are forbidding everyone to use? Most Malagasy people living in the region have little idea why the local price of rosewood has exploded during the past two decades (while other wood prices have remained stagnant) and why only the *chinois* are interested in buying this particular wood. Likewise, they are not sure why *vazaha* so desperately want to protect this seemingly unremarkable tree. Why leave a tree that is worth US$60,000 per ton in timber markets in China standing thousands of miles away in the forests of Madagascar?

The concept of endangered species does not provide a particularly helpful explanation. In Malagasy, "endangered species" is often translated as *biby efa ho lany taranaka* ("an animal whose generation will soon run out"). It does not quite work for trees. When I mentioned to a number of my interlocutors the possibility that rosewood might indeed "run out," some simply laughed, insisting either *c'est une plante régénérée!* ("the tree grows back!") or *maniry fô!* ("it grows everywhere!"). A few individuals went so far as to suggest that the concept of endangered species was a total fabrication, one of an array of foreign inventions meant to achieve domination through fear. Labeling rosewood an endangered species, they explained, created the regulatory pretense for prohibiting the trade and confiscating existing timber stocks. For these individuals, stories of endangered species found their place among other stories from foreigners—including AIDS and birth control—that a surprising number of my interlocutors thought were created and disseminated as a technique of foreign control. The COVID-19 pandemic demonstrated the immediate suspicions of all things foreign most forcefully: many in northeastern Madagascar believed it to be an "elimination system" started by foreign billionaires, governments—you name it.

As with viruses and birth control, conservation inspires skepticism and a tendency toward conspiracy theories. Not buying the conservation narrative, a number of my Malagasy friends insist that the national parks protecting rosewood were established to facilitate the future extraction of resources. They tell me that foreign visitors are not engaging in leisurely strolls but rather are scouting for resources—timber, gold, precious stones. The possibility that these foreigners are "just looking" at trees and animals appears to them to be a poorly crafted guise.[14] The forest is forbidden (*fady*), many Malagasy residents insist, only so that the foreigners themselves might use it in the future. This scenario, one friend explained to me, is called "parallel financing" (*financement parallèle*): a situation in which one party takes advantage of another by offering a few trinkets in exchange for a much greater prize. He had learned this term in school and asserted that it described many foreign investments in Madagascar quite well. With regard to the national parks, he observed that Malagasy people were offered conservation assistance in the short term but were losing out in the long term. Yet when I inquired what exactly he thought we foreigners wanted in return for the parks, or what we would do with them in the future, he could not answer. "That," he replied, "is what I was hoping you would tell me."

Armed primarily with concepts of endangered species and pristine nature, I had a difficult time explaining exactly what I thought we foreigners wanted with the forest. He was clearly dissatisfied with my attempt, as were many Malagasy people with whom I dared to share the "endangered species rationale," as I have come to call it. Occasionally in our conversations, another seemingly more compelling explanation surfaced: that (to my great surprise) Americans want rosewood logs for themselves. A rosewood trader I met in a logging village phrased this conviction most bluntly. The old Malagasy government, she explained, had a contract for rosewood with the Chinese, but the *new* government (those voted into office in 2013) had broken the contract in favor of the Americans. The Chinese are angry, she continued, because the new Malagasy government wants to have a contract with the Americans instead. Next, she warned, gesturing toward me and my American colleague, it will be you Americans coming with your boats to buy rosewood from the forest.

On some level, this answer is not wrong. Madagascar sits at the intersection of two global demands for rosewood, both attempting to buy the favor of the Malagasy government. The equivalent of more than US$1 billion has been pumped into Madagascar from both sides—to finance either rosewood conservation or logging.[15] Chinese importers have allegedly bragged that their money "can go through even the most highly placed doors," and Western donors have created a special task force in an attempt to ensure that this does not happen.[16]

Yet rather than stopping or slowing the trade, conservation task force members merely impose fines as logs move on their way. "Precious timber is a resource like any other," one task force member said to justify his fining (rather than stopping) of the trade. "Everyone needs to get their slice of the cake."[17]

These interpretations of the significance of rosewood in Madagascar—that of the rosewood trader who claims Americans want rosewood for themselves and that of the task force member who is trying to get his "slice of the cake"—may not be technically wrong. They only miss the nuance of exactly who wants what and why. These interpretations miss the deeper reasons why an otherwise unremarkable tree in the forest has become such a desirable "cake" for others across the world. Chinese retailers point to the rarity of the wood as a reason to preserve it for centuries in the form of furniture, whereas conservationists see an undisturbed forest ecosystem as the only means of preservation. Just as one country's theft is another country's repatriation, approaches to endangered species differ according to cultural and geopolitical positioning.

"The point of ethnography," ethnographer Anna Tsing observes, "is to learn how to think about a situation together with one's informants."[18] This has been my goal in Madagascar, and later in China, with regard to rosewood. I did this first, unwittingly, as a Peace Corps volunteer as I saw the impacts of the trade unfold, and then later, quite deliberately, as a researcher. I returned to the same region of Madagascar to conduct ethnographic fieldwork in the summers of 2014 and 2015. I performed semistructured and completely unstructured interviews with rosewood loggers, traders, and conservation agents through conversations that often stretched over the course of days or weeks in an attempt to create an oral history of the trade in the region. Together with loggers and traders, I made maps of the logging trails, which stretched from the interior of the forest to the surrounding villages and finally to the coast for export. I also conversed daily with residents indirectly affected by the trade as a result of price inflation, increased military presence, and the seemingly bizarre spending practices of newly paid loggers and traders. To this day, I continue to maintain contact over social media with residents living in the region.

My research in Madagascar was an attempt to construct a local perspective of the rosewood trade—to think with my informants—in order to supplement existing international accounts. Although I maintain an ethnographic approach, my ultimate goal is to understand rosewood as a global phenomenon. Acknowledging that there is no abstract layer enveloping the planet that one might call the "global," I approach global dynamics as (dis)connections across time and space that bring disparate cultural, political, economic, and ecological elements together, even if only momentarily. This approach, rooted in critical geography, political ecology, and cultural anthropology, is open ended more than systematic. "Research categories develop *with* the research, not before it," Tsing observes of ethnography. This indeed has been my experience: my most foundational understanding of rosewood—along with the concept of endangered species more broadly—has drastically transformed since I first landed in Madagascar more than a decade ago. This transformative journey is what I hope to recreate in this book.

Although the inspiration for this book began in Madagascar, it was in China that the story came into its full form. Since 2005, I have been traveling to Shanghai every few years with my husband, who is Chinese and has an exceedingly gracious extended family who hosted our stays. There I witnessed an entirely different transformation from that of Madagascar. Within the span of a decade, old apartment buildings, markets, and makeshift slums were torn down and replaced by high-rises stretching to the horizon. I went from sticking out like a sore thumb—with locals randomly asking for pictures, besieging me to follow them to their counterfeit collections, relentlessly questioning my (eventual) husband whether we were in fact dating and whether he was really Chinese—to becoming entirely unnoticeable, camouflaged amid the modern cityscape. The neighborhoods I visited were "rising up" quite literally but also figuratively. In less than a decade, Shanghai transformed from a center of cheap Chinese knockoffs peddling European and American distinction to a center of elite Chinese consumption that put Western shopping sprees to shame.

This transformation, however, was not turning the Chinese into carbon copy Western consumers, as many outsiders anticipated. Indeed, China's consumer revival is as cultural as it is capitalist. A rich cultural history is being revived in China's contemporary embrace of capitalist modernity—one that, I came to understand, is aptly symbolized in the revival of classical rosewood furniture.

My trips to China were largely for fun and leisure until I started my fieldwork in 2014. The need to expand my fieldwork from Madagascar to China was prompted by a comment from my husband, who casually pointed out that I knew almost nothing about the demand for the wildlife I was studying. Even to call it "demand," he insisted, was a brutal reduction of the immense cultural and economic convolutions driving wildlife consumption in China. To be sure, I had a full arsenal of go-to explanations: Ming Dynasty furniture, exotic tastes, rapid economic growth, and the nouveau riche. But I did not grasp the gravity of what it all meant. Whatever causal chain I had erected in my head at the time was largely a product of the Western imaginary of China I had absorbed.[19] The vast landscape of reporters—even academics—covering the topic suffered from this weakness too.

Although Western media outlets interpret Chinese demand for rosewood and other endangered species in terms of "exotic tastes" and "conspicuous consumption," the reality of the situation goes much deeper. Indeed, the importance of rosewood in China at this particular moment in history cannot be overstated. Rosewood looms large in China's material reality and collective imaginary, representing not a niche cultural artifact but one of the most powerful icons of the rise of the country. It is in fact difficult for those I interview to come up with a household investment that more aptly symbolizes Chinese identity ("A rice cooker?" one respondent suggested in partial jest). From the perspective of those of a certain age living in China, the story of rosewood—crafted for emperors centuries ago, violently confiscated from families within living memory during the Cultural Revolution, and now selling for millions in Chinese timber markets—is quite simply the story of China. Its ups and downs—its original grandeur, momentary devaluation, and contemporary revitalization—mirror those of the country writ large.

Remarkably unaware of all this, the Western popular imaginary sees rosewood—if at all—as confirming familiar narratives of China's rapacious demand for natural resources. Whether one has heard of rosewood or not, one has surely heard the recurring narrative that begins with the tragic loss of endangered species globally—elephants, rhinoceroses, tigers—and ends with burgeoning demand from a growing Chinese elite. The narrative invariably channels Western anxieties about the growing economic and geopolitical power of Asia. I recall attending a lecture by a prominent conservationist who, after documenting a rise in elephant poaching, remarked that what was driving this increased poaching was simply conspicuous consumption—not even traditional medicine, not anything *needed* by anyone, just rich people showing they are rich. This was likely not a revelation to the audience. Indeed, this is the understanding of many people who work on the topic, myself included at the time.

Conducting research in China has changed my perspective. Following my fieldwork stints in Madagascar, I traveled to China, primarily to Shanghai and Beijing during the winters of 2014, 2015, and 2017, and Guangdong Province during the spring of 2018. During this time, I interviewed timber importers, furniture makers, and urban Chinese families who had recently purchased rosewood furniture or who maintained a family heirloom. I visited Shanghai's main wholesale timber markets, rosewood manufacturing sites, and furniture retailers and expositions. I also visited three rosewood plantations to inform my analysis in Chapter 6. Supplementing my fieldwork, I reviewed market reports on the rosewood trade in China since 2000 and Chinese policies governing the trade. In order to historically situate this contemporary demand, I analyzed the cultural history of rosewood in China dating back to the Ming and Qing Dynasties. This historical analysis involved secondary textual reviews of artisanship during the Ming and Qing Dynasties and interviews with individuals and families who lived through China's Cultural Revolution and who, in many cases, had their rosewood family heirlooms confiscated or destroyed.

As with my research in Madagascar, my research in China adds new perspectives to the debate over rosewood, including the trade in endangered species more broadly. There are a thousand stories one might tell about Chinese demand for endangered species, yet we in the West always seem to hear the same one. Rooted in ethnography, my research in China attempts to tell a different story. My goal is not to romanticize or reify the appeal of rosewood in China—even less to legitimate it—but to provide a better understanding of how the resource is valued by those who came of age during the Cultural Revolution and who have experienced unprecedented wealth since. I do this not because I believe the pillaging of precious timber from the forest is justified but because I believe the drivers of the trade are deeply misunderstood. The consequence of this misunderstanding is that demand for endangered species is approached as a condition to be remedied—a symptom of a larger disease—rather than a complex intersection of culture and economy to be examined in its own right. Stepping away from the familiar narrative of Chinese demand for endangered species—provincializing it, so to speak—frees us to examine the phenomenon without an overarching policy objective. Ironically, it is this step away from analyzing the trade through a prescriptive lens that will help make room for fresh approaches to mitigating the trade that have a better chance for success.

Concerning rosewood, for example, the inevitable policy prescription is restrictions on logging and trade—an approach that is surprisingly *welcomed* by many Chinese timber traders and furniture owners with whom I have spoken. The prohibition on trade in rosewood from Madagascar in 2013 was in fact a great boon to the market, sending upward shocks in what was later deemed the "Madagascar phenomenon" (*Mădájiāsījiā xiànxiàng*)—that is, the tendency for trade restrictions to trigger market booms. When I told my uncle-in-law in Shanghai—a proud rosewood furniture owner—about all this, he grew excited. Noticing my confusion, he explained that this should make his own collection of African rosewood much more valuable. When I pointed out the sobering reality that mature rosewood will likely soon be gone from the forests of Madagascar entirely, it was his turn to be confused. "Why is this a problem?" he inquired. By making their way to China, my uncle-in-law assured me, the trees are in fact realizing their superior form. Their beautiful rose-colored cores will come alive in the form of classical furniture inspired by artisans of centuries ago and to be passed on for generations to come. As far as he was concerned, it was a win-win.

two trails in the forest

The value of rosewood is clearly in the eye of the beholder. All actors involved with this global resource—conservationists predominantly from the United States and Europe, timber importers and consumers from China, and residents of northeastern Madagascar navigating the global junction—lay different claims to the tree. For the conservationist, the fallen tropical hardwood symbolizes the relentless onslaught against one of the most biodiverse ecosystems in the world. The fight against rosewood logging is symbolic of the larger fight against biodiversity loss in the tropics. For the modern Chinese consumer, classical rosewood furniture symbolizes the cultural sophistication of a nation yet to be properly recognized by the global community. The acquisition of rosewood furniture has become a pillar of Chinese identity. Both imaginaries, of rosewood as endangered species and of rosewood as cultural icon, dictate the tree's worth in an increasingly contentious global arena.

In Madagascar, these divergent global demands manifest in the forest as two trails that run side by side but never meet. There are tourist trails and there are logging trails, each specifically designed to avoid the other. Rangers warn new guides not to let tourists venture onto logging trails, and logging bosses likewise insist that their men stay off the tourist trails. In Masoala National Park—the site of the greatest logging effort—tourists often ask to see endangered rosewood trees, having heard about their plunder at the hands of the Chinese. Appeasing the tourists, guides obligingly pass by a solitary tree growing at the base of one of the trails, thirty years old and still too small for commercial exploitation. The tree—covered in thick bark that hides its splendid rose-colored core—is unremarkable. Tourists nod, recognizing the species to have sparked sharp international controversy despite its meager appearance, and continue along the trail. Meanwhile, on the logging trails not too far away, century-old rosewood trees are felled, with bright red splinters scattered at the stumps. Giant logs are dragged through forest and river in dangerous feats that often leave loggers injured or dead. Eventually, as the logs reach the coast, they are delivered to unauthorized ships floating at the horizon and begin their journey overseas.

Following each type of trail in the forest reveals a different value of the tree. Along the tourist trail, one sees the value of the tree in the forest—an endangered species surviving at a density of at most one to two trees per hundred hectares, a rare node within its broader forest ecology. Along the logging trail, in contrast, one sees the value of the wood apart from the forest—the beautiful rose-colored core that will eventually be sculpted into ornate classical furniture originally designed centuries ago and made to last for centuries to come. Far from being killed, rosewood quite literally comes alive in the process of its crafting, demonstrating the progressive transformation of forest into furnishing. From these opposing positions, there is no ground to be given; there is no compromise possible. The tenets of one position (to conserve rosewood trees as endangered species) preclude the tenets of the other (to transform rosewood trees into their superior crafted form). They cannot both be realized at the same time for the same trees.

Yet when one chips away at these global imaginaries and examines their area of overlap more closely in all its specificity, the opposition begins to unravel. These opposing perspectives in fact meet at the moment when the first few blows of an axe expose the bright red heartwood of a towering tree in the forest. For this brief moment, in which the tree remains standing yet exposed, the two distant values of the tree—for the first time—stand side by side in a discomforting cohesion. A part of the forest is killed, but the hidden inner beauty of an otherwise ordinary tree is brought to life. Surely neither is watching at this point in the process, but if they were, the conservationist could not deny the beauty of the tree only now exposed, and the furniture connoisseur could not ignore the beauty of the forest in which it stands. This meeting point of tree and forest reveals a hidden junction where both sides of the story converge and both values occupy the same space. This is a space of contestation but also a space of common ground.

Rosewood, as a conflicted global resource, tells us exactly what we expect to hear about natural resource extraction and struggles over land: East versus West battling it out in the resource-rich "Third World." But if we listen very carefully, it also tells us the opposite. Like chipping away at thick bark and sapwood and exposing the red heartwood underneath, the full story of rosewood reveals both the stereotypes of natural resource struggles and their unique foils. It engages with the realities of conservation science as well as the mythologies that underpin them, the rationalities of capitalism as well as the cultural logics they articulate, and the global blind spots—such as the forests of Madagascar—that have paradoxically become quite central to ongoing land and resource competitions.

This book will show both sides of the story. At times, the United States may seem to be a harbinger of conservation, development, and democracy; at other times, a harbinger of mythology and mysticism. China will resemble both a leading environmental threat and a potential environmental leader. And the remote forests of Madagascar will appear increasingly central to global resource disputes. Conventional wisdom maintains that in order to understand globalization we must look to the most interconnected, cosmopolitan centers, but this book will show precisely the opposite. Following the path of rosewood as it travels from the forests of Madagascar overseas to the timber markets of China, readers will see how peripheral lands that seem least connected with metropolitan hubs of development reveal global dynamics in profound ways. As Chinese demand for rosewood spikes to unprecedented highs and Western conservationists fight to stop the logging, the remote forests of Madagascar have become one of the most global places rather than the least. Understanding such spaces where conflicting global demands meet is key to understanding their strained opposition. Too complacent with one side, we miss the rose-colored core of the other. Too focused on the clash, we miss the common ground.

reinventing the past

Global demands to conserve and commoditize rosewood reveal fundamental symmetries. Conservationists aspire to preserve remaining rosewood territories across Madagascar as pristine remnants of a ravaged globe. Meanwhile, in the homes of China's new consumer class, rosewood furniture provides a daily reminder of a romanticized national past for a tumultuous generation. Both approaches to rosewood follow a preservationist logic that reinvents an idyllic past to meet the needs of the present. The difference is what exactly is being reinvented—nature or culture—and by whom.

The preservationist logic that underscores global conservation efforts is well explored. Scholars critical of strict nature preservation question the existence of "nature" as an objective category in isolation from the human.[20] Nature itself, they argue, is a profoundly human creation, a "reflection of our own unexamined longings and desires," which "could hardly be contaminated by the very stuff of which it is made."[21] Protecting pristine nature from humans is therefore tantamount to protecting the created from the whimsy of their very own creators; it ultimately makes no sense. According to these scholars, the ability—and desirability—of a Western conservation agenda to preserve nature as separate from the human world also makes little sense.

The preservationist logic that underscores China's demand for endangered species is, in contrast, less explored—with the exception of tokenized explanations of exotic tastes and burgeoning wealth that serve more to vilify than to illuminate. Again, my goal in forging this exploration is not to legitimate a trade that contributes to the decline of endangered species but rather to foster a better understanding of it. As with classical art, China's renewed demand for endangered species is a cultural revival based on a historical legacy. Rosewood furniture in many ways represents the material embodiment of this shared legacy. Beyond rosewood, Chinese art repatriation and demand for endangered species more broadly are also properly understood only within the context of this wider historical narrative. Earlier, I sketched it in broad strokes: imperial opulence, the twentieth-century fall from grace, and the twenty-first-century revival. In the pages that follow, I add more flesh to this historical narrative, which conservationists who emphasize the ecological importance of rosewood and other endangered species hardly know.

The story of rosewood in China—like the story of the rhino horn goblet and the other antiques "stolen" in the art heists that have plagued the past decade—began centuries ago in imperial China. For rosewood furniture making in particular, this esteemed cultural work has humble origins. The craft originated in rural, largely impoverished areas in southern China where tropical hardwoods grew in abundance. With few other means of making a living, rural villagers attempted to extract as much value out of their woodworking as possible. They developed elaborate designs, deep polishes, and clever joinery requiring no nails or screws to hold their furniture together (Figure i.2). They sold their products not to one another but to the elite in neighboring cities. At first, this type of hardwood furniture was considered an inferior substitute for similarly styled lacquered furniture. But as the craftsmanship evolved, the style of furniture ascended the social ladder, eventually coming to dominate the royal scene.

During the late Ming Dynasty and well into the Qing, emperors commissioned the production of rosewood furniture, advocated the craft as a mark of cultural sophistication, and even participated in the work themselves. Trickling down through these elite capillaries, the furniture style increasingly became associated with wealthy social elite and the Chinese imperial literati. "One only needed to see the furniture in a family's main hall," a Ming scholar observes, "to appreciate the household's social position, economic power, and cultural level."[22] Furniture making became so revered, in fact, that the furniture itself was considered "to have a soul, epitomizing the cultural or even moral height of its designer and the taste of its user."[23] The arduous process of chiseling raw logs into exquisite furnishings was thought to liberate the soul contained in each piece of wood, freeing it within the homes of the imperial elite.

Rosewood furnishings in drawing rooms and studies thus represented not only the owners' wealth but also their cultural sophistication. "To change vulgarity

Figure i.2. A panel of rosewood decorated with the fortuitous dragon image. © George Zhu.

to elegance," a common saying of the time observed, "is as difficult as turning metal to gold; only one with the ability as vast as the hill or forest may achieve the transformation."[24] Indeed, forests across China were systematically decimated in the increasingly popular pursuit of transforming vulgarity into elegance via the acquisition of rosewood furniture. As with elephants, rhinoceroses, and tigers, whose populations were either extinct or greatly reduced in China by the early fifteenth century, precious hardwoods experienced early endangerment. Beginning as early as the fourteenth century, hardwood reserves were in decline as a result of imperial wood production. As the country opened to international trade with the installation of a new emperor in the early fifteenth century, tropical hardwoods from Southeast Asia destined for the imperial wood workshops constituted a primary import. Vast expeditions led by the famous explorer Zheng He brought Chinese ships to the shores of Madagascar and East Africa. Many of these ships came back with rosewood—including rosewood from Madagascar—to supply the imperial workshops. Following this, the dynasty's relative closure and "anti-maritime attitude" from the mid-fifteenth to mid-sixteenth centuries placed renewed pressure on China's hardwoods, causing near extinction for a number of species.[25]

The overexploitation of hardwood reserves both in China and throughout the trading region lasted well into the Qing Dynasty. Emperor Qianlong, for example, almost single-handedly caused the extinction of a once abundant local hardwood species (*nán mù*) in constructing his ironically christened Hall of Simplicity and Sincerity in the imperial palace during the mid-eighteenth century.[26] Previously taxed heavily, the harvesting of many hardwood species was by this time completely monopolized by the emperor. Imports of hardwoods from across Southeast Asia, referred to generically as *hóng mù* (literally "red wood," the common name for rosewood in China today), continued to replace domestic production of more traditional species, such as *zǐtán* and *huáng huālí.*

Rosewood's imperial legacy—as with most stories of Chinese tradition—ends with the fall of the Qing Dynasty. In part a result of domestic disorder and a depleted national treasury, and in part caused by rapidly declining hardwood reserves throughout the region, all aspects of furniture making suffered. Rosewood furnishings retained their iconic value, but manufacture ceased. Instead of new production, family heirlooms were passed down to new generations. In the decades that followed, wealthy families continued to boast elaborate displays, and poorer families sharing crowded units maintained small rosewood furnishings slotted between cramped mattresses and the chamber pot. By the mid-twentieth century, however, this would all change.

Beginning in 1949, China's Communist revolution initiated a nearly three-decade campaign of abolishing class distinctions and drastically reinventing the economic and cultural foundations of the country. This social upheaval is rarely discussed when touting the imperial history of rosewood furniture and other classical icons, yet it is essential to any understanding of their contemporary revival. The Communist Party, led by Mao Zedong, ushered in a series of campaigns aimed at radically redistributing wealth and prestige from the hands of the former elite. The movement culminated in the country's decade-long Cultural Revolution from 1966 to 1976. After centuries of valorization, rosewood and other traditional icons were abruptly devalued. Classical rosewood implements, rhino horn goblets, ivory decorations, ancient porcelains, and so forth were suddenly considered wanton emblems of bourgeois oppression and public scorn. Revolutionary groups, such as Mao's rising Red Guard, were encouraged to ransack the homes of former elite in search of these profligate possessions. If found, they were confiscated or destroyed, and the harboring families were paraded around the neighborhood, beaten, and scorned.[27]

The campaign to revolutionize "bourgeois tastes" hit the rosewood industry especially hard. During this period, the dynastic craft of woodworking was entirely forbidden. The price of rosewood plummeted as wealthy families hurried to jettison their rosewood possessions before Mao's Red Guard came knocking at their door. Priceless family heirlooms were surrendered for a small fraction of their worth. Consignment stores swelled with these and other cultural goods, now almost worthless. Confiscated furniture was either burned in large demonstrations or thrown into warehouses for later redistribution. After the Cultural Revolution, it was not uncommon to find antique rosewood furniture worth millions in today's markets serving as chopping blocks for rural farmers.[28]

But the drastic devaluation of traditional icons was only temporary. By the time of Mao's death in 1976, the Cultural Revolution ended and, with it, Mao's social experimentation. The Communist Party took a reformist turn, denouncing the former program of eliminating class distinctions and embracing an entirely new program of economic development. Those who managed to hold on to their rosewood furniture—or who purchased it for cheap from a consignment shop during the revolution—would once again see the cultural and economic value of the wood grow. "To get rich," the Party's new reformist leader, Deng Xiaoping, is said to have declared, "is glorious."[29] Just as in Ming and Qing consumer society, rosewood was to play a prominent role in the newfound wealth.

With China's reformist turn, many families that had endured the hardships of revolution—assault, dislocation, food rationing—were by the late 1990s greeted with unprecedented wealth. After decades of cultural asceticism, a new "luxury China" emerged, replete with a growing consumer class increasingly beholden to the demands of capitalist consumption.[30] Decades of social leveling campaigns had left citizens with little heritage of class distinction and, according to some, "desperately seeking status."[31] As families tried to rebuild a legacy of class distinction, or evade the stereotype of the uncultured nouveau riche (*bàofā hù*, as they are referred to in Mandarin), cultural goods played a decisive role. Classical icons formerly scorned by the Communist Party came to feature prominently within the Chinese imaginary as a source of national pride. All across China and its diaspora, Chinese traditional culture experienced an economic reinvigoration. In the context of this mounting traditional reverie alongside capitalist integration, markets for rosewood and other cultural goods boomed.

Thus far, the consumption of rosewood and other endangered species has been propelled largely by the older generation, those who lived through the chaos of the Cultural Revolution and Mao-era social upheaval, food rationing, and famine. Chinese youth who have known only an economically open and prosperous post-Mao era have an entirely different experience of what it means to be Chinese. The generational gap that China now faces is perhaps the largest ever to exist. "Parents who spent their own twenties laboring on remote farms," an essayist notes of this gap, "have children who measure their world in malls, iPhones, and casual dates."[32] It remains to be seen whether Chinese millennials will embrace rosewood and other culturally important endangered species at the scale their parents do. Yet, as this book will consistently demonstrate, the complete Westernization of Chinese consumer tastes, social structures, political dispositions, and approaches to environmental reform is highly unlikely.

Beyond its traditional associations, rosewood has been increasingly purchased purely for investment. As the Chinese nouveau riche began to buy back their lost cultural heritage after the Dengist reform period, investors paid close attention. Not only were rosewood prices likely to increase because of renewed consumer demand, but also the materiality of the resource—the scarcity of the wood in the forest—ensured that prices would soar even higher. The unique combination of scarcity, durability, and cultural appeal, as discussed in Chapter 1, has made rosewood one of the foremost speculative investments in contemporary China. By 2005, the rosewood market had become "a playground for investors," catering not to consumers, connoisseurs, or collectors but to financiers looking for quick returns. The speculation extends beyond rosewood to many other high-value commodities derived from endangered species, such as ivory, rhino horn, tiger parts, and pangolin scales.

China's modern conjuncture makes sense only in light of the history that came before it. There is a unifying perception, especially among survivors of the Cultural Revolution, that amid the chaos of social upheaval and global economic integration there are few other stable and legitimate avenues for surplus capital investment. Those who remember the Cultural Revolution remember years during which money was almost worthless, wages were paid in stamps or vouchers for food and other necessities, and the state owned all property. To go from such a collectivist reality to the late capitalist efflorescence of contemporary China in a single lifetime is to know a kind of social and economic vertigo that leaves one permanently leery of what others might characterize as solid ground. Financial institutions such as banks and bond markets could disappear overnight; publicly traded private corporations could be reintegrated into state-owned monopolies; the real estate market could be (and often is) manipulated at the whim of state policy. For many Chinese consumers, investment in art, antiques, and endangered species represents an acknowledgment of one of the few indisputable facts of contemporary Chinese society: as long as there are Chinese people, there will be demand for the cultural goods that play a crucial part in how they define their identity.

investing in the future

The Chinese art thefts / repatriations of the past decade, along with renewed demand for endangered cultural resources such as rosewood, are examples of the resurgence of traditional Chinese culture in the global arena. This turn toward investment in cultural goods is, of course, not new—neither for China nor for the world. Art has long been used as a vehicle for storing and accumulating economic value via cultural prestige. Yet the contemporary dynamic in China demonstrates this trend with surprising intensity. Cultural goods that were practically worthless mere decades ago are now in some cases worth more than their weight in gold. Not only antiques but also a US$26 billion industry of newly manufactured rosewood furniture have become focal points of speculation. A resource that was once burned in the street or pawned for almost nothing is now driving importers to the farthest corners of the world.

For endangered species protection, the scenario seems devastating. New demand from China undermines Western conservation efforts in a predictable global pattern. As Chinese consumers buy back their lost cultural heritage in the form of rosewood and other endangered species, global conservationists fight these burgeoning demands with little understanding of what fuels them. This has resulted in unintended consequences. Rosewood conservation efforts, particularly international trade restrictions, can trigger market booms, as discussed in Chapter 5. Trade restrictions ensure the rarity of the wood, exaggerating the scarcity of supply and artificially driving up the price. This counterintuitive phenomenon of international trade restrictions boosting market prices—what Chinese timber traders now refer to as the Madagascar phenomenon, as noted earlier—means that conservation efforts can end up being a boon for markets in endangered species. It is no coincidence, for example, that rosewood prices spiked in 2013 after a number of trade prohibitions went into effect and that annual market sales peaked the following year, exceeding US$25 billion in 2014.[33] Contrary to their intended purpose, conservation restrictions have played a significant role in these market dynamics.[34]

Trade prohibitions and criminalization are the most conventional routes for conserving rosewood and many endangered species, but there are other options. China is in fact pursuing some of these alternatives. As discussed in Chapter 6, beyond simply exploiting rosewood, the national government and private investors are also conserving it—but in ways that differ from the Western precedent. Together, research institutes and intrepid entrepreneurs are planting hundreds of thousands of hectares of rosewood forests across the country as a financial, cultural, and ecological investment (Figure i.3). Discouraging eucalyptus and other fast-growing species, government agencies now promote the planting of rosewood and other slower-growing species across southern China, where climatic conditions are conducive to tropical hardwood growth. Entrepreneurs looking for new business ventures are following suit. This model of plantation agriculture is also being exported to neighboring Cambodia and Laos. Madagascar would surely benefit from such ventures as well, given the opportunity.

Rather than exaggerating the scarcity of the wood through trade restrictions and logging prohibitions—that is, rather than focusing on *de*forestation—conservation policies might be more effective in focusing on *re*forestation and other sustainable forestry activities. Establishment of rosewood plantations offers a promising alternative to help preserve species and meet future demand. Some wealthy residents of northeastern Madagascar have already begun planting mixed agroforestry gardens with rosewood as a type of inheritance for their children. Yet conservation groups that influence policies in Madagascar—for no apparent reason other than their prioritizing of pristine nature over a working landscape—overwhelmingly emphasize logging prohibitions rather than policies that promote planting. This often results in a type of militarized

Figure i.3. A government plantation of rosewood and other precious hardwoods covering more than two thousand hectares in Guangdong Province, China. The plantation is a demonstration plot meant to encourage the planting of noneucalyptus species.

conservation that, as we will see in Chapter 4, does little to actually protect trees from being felled and even less to meet the needs of those living closest to the forest.

Debates over how to conserve rosewood—whether to preserve the last remaining trees or to plant the species anew on a large scale—are indicative of a deeper divide in global battles over endangered species. This divide concerns not just material resources but also visions of a sustainable future. Whereas Western countries advocate trade bans and nature protection, China promotes reforestation and sustainable use. China is in fact the only country planting rosewood on a large scale, not just for biodiversity conservation—and even less for nature preservation—but rather to invest in a future that supports the country's unique cultural, ecological, and economic aspirations. In this sense, China's model for growing rosewood as part of a working landscape clashes with Western efforts to conserve the species as pristine nature. Each side embraces a different vision of a sustainable future when it comes to rosewood. This conflict plays out in the tropical forests of Madagascar as well as in the halls of global diplomacy, where policies for the protection and use of wildlife are hotly contested.

There is a growing environmentalism in China, but it is not the environmentalism of the West. This is a vital, and often overlooked, point. Although differences between Chinese and Western approaches to the environment mattered little before, they have become pivotal in shaping international environmental efforts now that China exerts such force on the world stage. In an attempt to decenter Western environmentalism and look toward the future, this books asks what such non-Western environmentalisms look like. How might rosewood show us not only China's voracious demand for resources but also its fight to preserve these resources in the coming century? Global environmentalism of the twenty-first century—and any hope of combating biodiversity loss—will necessarily emerge through negotiating and reconciling Chinese and Western approaches to the environment. Our shared global environmental future depends on collaboration with China; it cannot be achieved alone. Such collaboration, however, requires first a sincere understanding of China's approach to endangered species and the environment.

theories and debates

This book builds on foundational theories in the fields of political ecology, critical geography, and cultural anthropology that question the universality of concepts such as nature, wilderness, and endangered species. Rather than objective and universal categories, nature and associated concepts are understood in this book to be social products, or socially constructed categories. Instead of nature, political ecologists use the concept of "socionature," which emphasizes the deep entanglements between observations of the natural world and the cultural dispositions of the observers.[35] According to political ecologist Nancy Peluso, a socionature is a "Latourian hybrid between cultural conceptions and representations of nature and the actual 'natural objects' and social relations they describe."[36] The term is intended to denaturalize nature, so to speak, complicating nature / culture binaries that so deeply inform Western environmental thinking. The result is the revelation that nature is "inescapably social" and a potent breeding ground for political conflicts and power struggles.[37]

The implications of a denaturalized nature for conservation are profound. By questioning the existence of a pristine, humanless nature, critical approaches in the social sciences problematize preservationist models. They cast doubt over the mission of protecting nature from humanity and cast further doubt over the scientific concepts generated in pursuit of this mission. Biodiversity, for example, emerged as a scientific category in late-twentieth-century America in an attempt to give this cultural conception of nature objective rigor.[38] In *The Idea of Biodiversity: Philosophies of Paradise,* environmental law scholar David Takacs outlines this emergence.[39] Biodiversity, he reflects, transforms vague ideas of nature into the specific players of genes, species, and ecosystems interacting on an eco-evolutionary stage. Endangered species, too, perform a similar function. As environmental historian William Cronon notes, endangered species "stand as surrogates for wilderness itself," and those places harboring the last remaining populations, such as lush tropical rain forests, become "the most powerful modern icon for unfallen, sacred land."[40]

Assimilating a particularly American understanding of the natural world into the world of science is not in itself a problem, but it becomes one when sent overseas under the guise of purely objective science. Biodiversity loss is a problem of global proportions, and by exporting biodiversity loss solutions to lower latitudes, conservationists often view themselves as "healers of broken places."[41] They typically fail to notice, however, that they are exporting not only a rigorous science but also the situated ethic of protecting the nonhuman world—a category that does not exist for most of the world's people.[42] "Exporting American notions of wilderness," Cronon concludes, "can become an unthinkable and self-defeating form of cultural imperialism."[43] To proceed as if biodiversity were an unquestionably global issue rather than a highly situated manifestation of one particular understanding of the nature / culture divide would be to confuse, as people often do, science with its cultural underpinnings.[44] It would be to deny, as writer and historian Ramachandra Guha notes, "the cultural rootedness of a philosophy that likes to present itself in universal terms."[45]

This is precisely what has happened in Madagascar. Containing "more genetic information per surface unit than any other place in the world," Madagascar provides a vital landscape for transforming conservation's Edenic dreams into scientific realities.[46] The country's remaining forest corridors have been designated one of the world's top biodiversity hotspots—a highly scientific designation based on an ultimately cultural category.[47] Lemurs, periwinkles, chameleons, and now rosewood give substance to the nature the West is intent on conserving. Conservationists delimit parks in Madagascar to optimize species richness, but more than this, they reveal a present and peculiar desire to preserve a timeless tropical past.

Analyzing rosewood as a socionature, as this book aims to do, means cutting against the grain, transecting the nature / culture divide that dominates global imaginaries of the tree. It means understanding rosewood not just as an endangered species but also as a symbol of a tropical landscape that gives substance to the pristine nature conservationists are intent on preserving. Outside of the conservation context, it means understanding rosewood not just as an exquisite building material but also as a material that gives substance to the cultural heritage that many in China are equally intent on preserving. And, most important, it means questioning the apparent opposition between these natural and cultural framings. Although rosewood might at first appear rather one-sidedly natural or cultural, a fresh analytical slicing reveals the tree's deeply socionatural entanglements.

What do the unique Western cultural foundations of wilderness, biodiversity, and endangered species mean for understanding environmentalism in China? In addition to examining the influence of Western conservation in Madagascar, this book examines the global rise of an alternative environmentalism not based so strongly on Western precepts. Using rosewood and other endangered species as an example, the book shows that Chinese approaches to the environment do not follow the same nature / culture divide that guides Western approaches.[48] China exploits rosewood in the name of its cultural heritage, but it also fosters the growth and survival of the species through plantation agriculture for the same reasons. Beyond rosewood and other endangered wildlife, Chinese environmentalism more broadly is informed by a different cultural context—and is thus often misunderstood in the West. As I argue in Chapter 6, China's approach to the environment is historically rooted not in the Western nature / culture opposition but rather in a philosophical tradition that does not so sharply delineate nature from culture. Thus, when it comes to the idealism and rhetoric that drive environmentalism, in China the focus is more on cultivating a productive harmony between humans and their wider surroundings than on preserving wilderness apart from them.

Of course, we must be careful when relying on such generalizations—Western, Chinese, Malagasy. These categories obscure a great degree of heterogeneity. They are examples of what sociologist Ching Kwan Lee observes as analytical constructs that conjure some degree of "false homogenization" to allow for broader comparison and interpretation.[49] Generalizations such as these can be both useful and problematic, but they are not to be confused with *essentializations*—that is, the attribution of a key explanatory factor, or "essence," that is then assumed to define these categories, to explain their inherent qualities. This book acknowledges certain distinguishing features that arise when bracketing off the world in such unwieldy categories, but it does not assume these features to be essential or eternal. Instead, they are contingent and provisional, stemming from the drastically different contexts that have given rise to the vast material-semiotic assemblages we now call the West, China, and Madagascar. Just as with art, cuisine, and language, these regions follow different *environmentalisms*—that is, different understandings of the proper place of humans within their wider surroundings. At least, this is my contention, discussed further in Chapter 6.

In highlighting the divergent philosophical and historical underpinnings of various environmentalisms, especially emerging movements in China, *Rosewood* contributes to long-standing debates concerning China and the environment. From Lester Brown's inaugural question in *Who Will Feed China?* to more recent titles asking *Will China Save the Planet?* and pronouncing that *China Goes Green*, debates over the country's environmental impact have seen a clear shift.[50] As China's global resource exploitation continues to be a pressing point of concern, a new awareness of China as a potential environmental leader has emerged as well. Yet China's environmental rise is riddled with doubts over whether the country's environmental reforms are more strategic than genuine.[51] Suspicions arise: Are these reforms merely a consequence of international pressure or a guise for consolidating political control? The debate consistently evokes notions of "authoritarian environmentalism"—or, as Yifei Li and Judith Shapiro suggest, "environmental authoritarianism," that is, "authoritarianism in green clothing."[52] Authoritarian *environmentalism* is a method of pursuing urgent environmental reforms through top-down, nonparticipatory approaches, whereas environmental *authoritarianism* reverses the arrows of causality. Rather than authoritarianism used out of environmental necessity, the environment is used as an excuse to further extend the political reach of the state.

While recognizing how frequently environmental reforms in China are used as a means to pursue a political agenda, this book aims to tell a different story. Since the early 2000s, there has been mounting pressure on Chinese leaders to address environmental problems: air pollution in urban centers, flooding in low-lying river basins, sandstorms and desertification in the north, water quality throughout the country. In addressing these problems, the Communist Party of China *is* further consolidating political control—not only because addressing environmental issues conveniently allows for other monitoring and / or coercive measures but also because it secures the party's legitimacy in the eyes of the people. By effectively responding to environmental concerns, those in power are securing their own legitimacy—both in the eyes of citizens and, to a certain extent, in the eyes of the world. The very act of solving environmental problems is by default a means of solidifying political control, simply by meeting the demands of citizens and gaining popular support. For the Communist Party, environmental reform is not an end in itself (nor is economic growth, for that matter) but is always subordinated to the larger goal of maintaining political control and social stability.

Yet beyond China's "authoritarian" approach lie other differences between Chinese and Western environmental approaches. China does not share Western liberal governance approaches but also does not share the preservationist ethic that inspires much of Western environmentalism. The nature / culture divide has not penetrated Chinese environmental discourse so thoroughly as in the West, and consequently the preservationist mentality has not either. In this light, China's approaches to endangered species (focusing more on sustainable use than preservation, as discussed in Chapter 5) and reforestation (focusing more on large-scale ecological engineering than preserving or recreating natural landscapes, as discussed in Chapter 6) take on new meaning. They provide an alternative vision of an ecological future that might be more appealing to groups around the world that also do not share a Western preoccupation with preservationism (such as the South African Development Community when it comes to rhino and elephant conservation, as noted in Chapter 5).

The final debate in which this book seeks to intervene is that on China in Africa and "global China" more broadly.[53] There is extensive literature on China's growing impacts in Africa and the world, in terms of both endangered species and the global timber supply chain.[54] Although such documentation has been extensive for decades, the latest plateau in the ongoing story of the rise of global China is the country's One Belt, One Road (*yīdài yīlù*) global vision, more commonly referred to as the Belt and Road Initiative. Through the Belt and Road Initiative, China has partnered with 140 countries to develop infrastructure and investment projects potentially totaling as much as US$8 trillion. The environmental impacts of such massive global reorientation will be profound. Considering the forestry sector alone, the Belt and Road is likely to result in significant global deforestation to make way for new railways, highways, and deepwater ports.[55] Just as in mainland China three decades prior, vast forests will be cleared in the process of establishing this new global infrastructure.[56] But also as on the mainland, vast forests will be *planted* as well. The Belt and Road will bring both environmental destruction and new environmental practices to member countries across the world. This is already beginning. Having constructed their own Great Green Wall, Chinese researchers are now assisting with Africa's transcontinental Great Green Wall across the Sahel. Along the Central Asian economic belt, another massive "green economic belt" made of poplar trees is being "built" as well.[57]

To the Western ear, all this *building* and *constructing* when it comes to the environment—"building a beautiful China" and "constructing an ecological civilization," as written into the Chinese constitution in 2012—smacks of greenwashing. It appears to cast projects actually intended for economic growth in an environmental light. Yet from a Chinese perspective, building and constructing on one hand and environmentalism on the other are not in such strident opposition. Even as critics question the sincerity of China's "green" efforts, both domestically and overseas, it is useful to recall that "greenness" is culturally construed. Building and constructing vast forestlands has been one of China's primary methods of rescuing the country from air pollution, flooding, desertification, and—it is hoped—global climate change. These feats are pursued not out of romantic or moral appeals to nature but out of practical necessity and aspirations for a balanced ecology. Massive tree-planting projects may not appear so "environmental" in the West, but they may look otherwise in other parts of the world.

Debates on global China—and especially China in Africa—ignore these differences. Instead, they primarily focus on the economic and geopolitical motivations behind Chinese interventions abroad. Again, the environment is seen as a pretense to extend control, but this time at the global level. China's increasing presence in Africa is typically framed in terms of a "new" scramble for resources, building on the colonial "scramble for Africa" more than a century prior.[58] The emphasis in both cases, new and old, is material: acquiring strategic resources for geopolitical control. This book shows another side to China's increasing presence in Africa too often overlooked. Struggles over resources are not only material battles but always—and perhaps even primarily—battles over meaning and representation. Rosewood—not at all utilitarian or strategic in the classic geopolitical sense but deeply culturally significant—demonstrates this. Rosewood is not a generic timber but a tree with rich histories based in divergent cultural configurations. Its particular biophysical characteristics—slow growth, deep hue, fine grain, rarity—gain new meaning as they combine with various cultural elements that define the tree as either an endangered species or a cultural commodity.

When China's expanding global reach is viewed through the lens of rosewood, then, different stakes emerge. China's geopolitical exploits on the African continent—indeed, across the globe—are not just renewed attempts at resource colonization but a new opportunity to reframe global spaces on non-Western terms. The subtler but more profound fight over the African continent is about whose doctrines—Chinese, Western, or otherwise—Africa will abide by. This applies to how resources are exploited but also how they are conserved. Whose vision of a sustainable future might Africa follow? Africa represents a complex intersection where different global visions are reconciled and built anew. Preservationism, sustainable use, and myriad other approaches to the environment all play out in African spaces. In the process, African people repurpose and reinvent Chinese and Western environmentalisms to meet their own material ends or to resonate more deeply with their own cultural terms. These manifold negotiations of resource use and conservation point to a deeper struggle within twenty-first-century geopolitics. Resource battles have as much to do with securing control as with building a sustainable future.

chapter overview

Chapter 1 begins in the largest rosewood commercial center in China, a historic town famous for rosewood since the Ming and Qing Dynasties. Beginning in the early 2000s, this town has experienced a revival. As China's middle class seeks to reconnect with its cultural roots, all things distinctly "Chinese" are experiencing a cultural renaissance, and traditional rosewood furniture is foremost among them. This chapter tells the story of rosewood's imperial tradition and contemporary revival, from its late imperial origins to its violent confiscations during the Cultural Revolution. The chapter culminates in the present investment boom. Individuals who endured the hardships of the revolution are buying back their lost cultural heritage, and financiers looking for novel investment avenues have taken note. It is this confluence of cultural legacy and rampant financial speculation, the chapter argues, that drives endangered species consumption in China today.

Chapter 2 shows the other side of China's rosewood craze as it reaches the remote forests of Madagascar. Far from the world's metropolitan centers, rural Malagasy villagers must nonetheless ride the most extreme waves of today's speculative economy. Occasionally overwhelmed by the sporadic inflows of cash from rosewood logging and other speculative export economies, villagers turn to "hot money" spending—the Malagasy vernacular deeply related to, but not to be confused with, hot money in the financial sense. Hot money in Madagascar refers to the tendency of workers to spend their earnings in fanciful sprees, from making all-night trips to the bar to leisurely pasting money on chameleons. The term is also used globally to refer to intensive financial speculation following short-term investment horizons—one of the primary causes of rosewood speculation in China. With little access to banking or credit and few productive outlets for channeling windfalls, China's hot money economy materializes in northeastern Madagascar haphazardly. Hot money in both Madagascar and China, the chapter argues, reveals the absurdities of money circulating in a capitalist system, in which money is simultaneously the key to fulfilling one's livelihood and something that must be tossed around hastily as if it were hot.

Moving from money to politics, Chapter 3 explores the political implications of the rosewood trade in Madagascar. Starting in the forest, the chapter shows how villagers have used rosewood logging as a way to "take back" the national parks that they consider to be in the hands of Western nongovernmental organizations (NGOs) and the state. Then, turning to the capital city, the chapter describes how a small group of rosewood exporters came to represent the region as members of Parliament. Once in office, these exporters monopolized the rosewood trade and destabilized the government, allegedly spreading "rebel money" to buy impeachment votes. Rosewood exporters have captured popular agitation from local constituents and redirected it toward the central regime in an effort to undermine the state from within. Paradoxically, the chapter concludes, those who wish to dismantle the state end up assuming some of its principal roles.

Chapter 4 shifts from rosewood logging to rosewood conservation, along with all its unintended consequences. The chapter demonstrates how conservation initiatives in Madagascar often backfire, increasing rather than decreasing logging activities. Funding in the hundreds of millions of dollars from the United Nations, the United States Agency for International Development (USAID), and the World Bank has been designated to set up a military-inspired "rosewood task force." Yet task force members see it as their job not to stop the trade but merely to impose fines on those who participate. Not being able to determine with certainty who in fact participates, authorities impose penalties on communities across the board. Under this "worst-case conservation" scenario, community members anticipate logging fines regardless of their individual adherence to the laws and head to the forest without hesitation to earn money by chopping down trees. In addition to contributing to this dynamic, US conservation groups try to fight it through community-based management and monitoring. Yet this type of surveillance conservation bears more risk than promise in the face of strong pressures for rosewood logging.

Chapter 5 turns to the demand for rosewood and other endangered species in China, demonstrating further unintended consequences of conventional conservation approaches. Not only rosewood furniture connoisseurs but also those interested in rosewood purely for its potential to appreciate in value are buying up rosewood across China and its diaspora. Beyond rosewood, other speculators hoard ivory tusks in central Africa or rhino horn in South Africa. Endangered species such as these have become highly speculative commodities. Within the context of this speculation, restricting the trade in "investment-grade" wildlife products artificially increases their scarcity and thus their value. International environmental restrictions intended to stop the trade can thus end up triggering market booms. Although this is common knowledge in Chinese timber markets, it is rarely discussed by Western conservation groups. Instead, environmentalists continue to push for tougher trade restrictions, and Chinese timber importers continue to reap the benefits.

Chapter 6 uses the case of rosewood to draw implications for global environmental movements more broadly. As Western conservationists fight the rosewood trade, a renewed practice of planting the species is sweeping across southern China. The Chinese government and a suite of private investors are turning to rosewood plantations as a unique form of ecological, cultural, and economic investment. The result has been hundreds of thousands of hectares of endangered rosewood trees planted across the southern half of the country. More than endless rows of trees waiting to grow, these agroforests mix long-term growth with shorter-term "understory economies" (*lín xià jīngjì*) in high-value cultural commodities (herbs, teas, essences). China's rosewood plantations, the chapter argues, represent a renewal of Chinese environmental efforts that circumvent a Western conservation ethic even as they attempt to save species. Including such unorthodox approaches to endangered species conservation, the chapter further maintains, can help "pluralize" environmentalism at the global level—ultimately reducing geopolitical tensions rather than heightening them.

The concluding chapter uses the case of rosewood to reflect on the broader stakes of China's global rise. Through rosewood and other endangered species, China brings new consumer demands, new cultural histories, and new visions of a sustainable future into the global arena. This is not to say that China's rosewood craze will spread internationally but rather that China's approach to endangered species and the environment more broadly—sustainable use, captive breeding, reforestation via large-scale plantation agriculture—very well could. China's global presence challenges Western norms, with consequences extending far beyond rosewood and endangered species protection. China's Belt and Road Initiative—also known as the "new silk road"—demonstrates this. In addition to reshaping the physical world, the Belt and Road unites "global Chinas" of past and present while glossing over the interstitial period of Western hegemony. What is truly at stake in the rise of global China, the book concludes, is how the country could rewrite globalization and the environment on its own terms. More than ever before, the West must reckon with divergent values placed on global resources as well as divergent approaches to their conservation. *Rosewood* shows how vastly different these new global approaches to endangered species and the environment can be, and how profound are the consequences for our shared environmental future.

❋ 1 ❋

cultural bloom, cultural boom

Although its population of three million would make it a major metropolis almost anywhere else in the world, the town of Zhongshan in southern China is considered a relatively quiet, provincial town of little consequence. That is, except for the town's famous rosewood industry. Hundreds of years ago, during the cultural zenith of the Ming Dynasty, the craft of rosewood furniture making swept across imperial China, with Zhongshan at its center. Today, Zhongshan is considered the rosewood capital of the world, with 80 percent of China's rosewood imports, I am told, processed in this district. This quaint town has become the center of a US$26 billion market for classical style furniture made of endangered precious hardwoods imported from across Asia, Africa, and South America.

Surveying the town in the spring of 2018 during a visit to its annual Rosewood Furniture Expo, my husband, my translator, and I were confronted by endless boulevards lined with rosewood dealerships. We had imagined the expo would be a modest affair, something akin to the niche trade shows hosted in convention centers of second- and third-tier cities across America. As we drove through the wide boulevards of Zhongshan, we realized we had been badly mistaken. Even three miles away from the expo center, we saw from our taxi window that every storefront, every business, every building was a rosewood furniture dealer. Their open doors lined the streets, beckoning potential buyers who had traveled from all over the country to step onto their thickly polished showroom floors and regard their elaborate collections. Some of the showrooms occupied entire blocks; some featured baroque entryways framed by enormous columns;

others advertised ornately carved chairs on three-story-tall billboards. The pièce de résistance was a five-story luxury shopping mall dedicated exclusively to the sale of rosewood furniture.

Upon pulling up to the front entrance of the expo center, we stepped out to see what appeared to be a classic Chinese temple but built at the scale of the Bellagio. This was no small-town convention center. This was a Chinese-torqued Vegas-style fever dream, a theme-park commercial palace dedicated to rosewood furniture. Containing nothing short of hundreds of rosewood dealers large and small, shopping arcades enclosed under an artificial sky, theater squares with live performances, a rosewood museum, and China's first university dedicated to rosewood, this mini rosewood-opolis facilitates the rapid exchange of hundreds of thousands of renminbi (RMB) from buyers across the country (Figure 1.1).

After climbing to the fifth floor of the main shopping center, we finally encountered the annual furniture expo, an event entirely eclipsed by the enormity of the "regular" business below. Displayed casually inside one of the better-appointed booths were two elegant chairs made of a waxy yellow wood called *Hăinán huáng huālí*. This highly prized rosewood is commercially extinct as fully mature trees, and as finished furniture made of the highest-quality wood, it can sell at a price per kilo above that of gold. Both chairs were in good condition, but when compared with the rest of the new furnishings on display, all of which were polished to a high shine, they had a noticeable patina. We were told that these chairs were made from an original Qing Dynasty wood stock from the personal collection of the owner of the company. Price? Too valuable to sell, the showman insisted, but if they were to have a price tag, it would amount to more than half a million US dollars for the pair. Antique furnishings made of the same wood have been known to sell for nearly US$11 million for a single chair.[1]

How did all this happen? What makes Zhongshan—the miles of dealers, manufacturers, and engravers; the palatial rosewood furniture exposition—possible? From a Western perspective, China's rosewood boom may appear wholly irrational, analogous to a sudden and inexplicable appreciation by middle-class Americans for Louis XIV armoires—or, at the very least, an overheated speculative bubble driven simply by greed. In addition to the decadent styles, the environmental impacts of the industry—the logging of endangered precious hardwoods across Asia, Africa, and South America—cast demand for rosewood in an indulgent light. Indeed, for many, rosewood reveals just another way exotic and arbitrary Chinese tastes manage to decimate global resources. Yet, as my trip to Zhongshan and travels throughout China more broadly have impressed upon me, we fundamentally misunderstand Chinese demand for rosewood and other endangered species.

Figure 1.1. A model of China's largest commercial center for rosewood and also the location of the annual Rosewood Furniture Expo, Zhongshan, China, March 2018. © Annah Lake Zhu.

China's rosewood boom, as evidenced by the sudden reinvigoration of Zhongshan and other townships historically dedicated to rosewood furniture production, parallels the country's boom in endangered ivory, rhino horn, and commodities made from other exotic species. Demand for these resources has emerged from a diverse assemblage of cultural and economic elements recontextualizing the contemporary Chinese milieu. Because of both their rarity and cultural appeal, rosewood and other endangered species offer a novel opportunity for speculative growth in an otherwise oversaturated investment economy. These markets provide a new cultural frontier for capital expansion. They absorb China's excess capital flows, artificially driving up the price of rosewood at the expense of the forest, with devastating effects for endangered species across the globe.

Rosewood in China thus provides the material nexus in which contemporary financial flows meet a revival in traditional culture dating back to the dynasties. It is impossible to understand the role of rosewood and other endangered species in contemporary Chinese society without capturing both their economic and cultural dimensions and, most important, the intersection between the two. Like any cultural identity, contemporary China's is based on a historical legacy. Whether factual or fictional, this legacy is shared, and thus it is recognized without explicit reference. Rosewood furniture in many ways represents the material embodiment of this shared legacy, which has become a major selling point in commercial Chinese folklore. Today, rosewood is not just an increasingly rare resource; it also happens to be one of the primary signifiers of Chinese identity. Displaying rosewood in one's foyer or study is not just about being rich but also about being *Chinese.* Chinese demand for rosewood and other endangered species—indeed, nearly all aspects of contemporary China—makes sense only within the context of this larger historical trajectory. It is precisely the confluence of this historical trajectory and global capital flows reaching the trillions that drives China's cultural boom in endangered species.

Using rosewood as a focal point, this chapter outlines the stages of China's cultural history that have been so influential in shaping the dynamics of the past two decades, starting with the fifteenth-century Ming Dynasty "cultural bloom," continuing with the mid-twentieth-century Cultural Revolution, and culminating in the country's new millennial cultural boom: a frenzied buying (and repatriation) of all things iconically "Chinese." Building on this history, Chapter 2 will show another side of China's cultural boom as it reaches all the way to Madagascar, resulting in "hot money" spending sprees across the rural countryside. But before moving to the supply side, we must first better understand the demand. The hot money of Madagascar's rosewood boom originates here in China: in a globally integrated, state-led market economy wherein Chinese cultural items represent one of the most promising investment avenues for a rising elite who endured the hardships of the Cultural Revolution.

beyond conspicuous consumption

China is by far the largest consumer of internationally traded endangered species worldwide. Data are limited in terms of exports and imports because most of these species are trafficked illegally. Seizure data of confiscated wildlife are, therefore, often used as a proxy. Rosewood is the most trafficked group of wildlife according to this data, accounting for roughly one-third of all seizures by value. Elephants are next, followed by pangolins and rhino horn (Figure 1.2). China is the overwhelming destination for all of these groups of wildlife, along with other Asian countries influenced by Chinese culture.[2]

These statistics likely come as little surprise to Western audiences. In the West, Chinese demand for endangered species is understood primarily in terms of conspicuous consumption: the Veblenian purchase of luxury goods as a form of social display.[3] Western audiences are familiar with what some have deemed "the imaginary 'Asian super consumer'"—a stereotype media outlets often conjure when portraying the wildlife trade.[4] The stereotype is on par with the "crazy rich Asians" mythos, portraying predominantly Chinese consumers with mounting purchasing power and a taste for exotic delicacies. "Chinese consumers' crazy rich demand for rosewood," an exemplary headline declares, "propels drive toward its extinction."[5] The stories rarely delve further into this mysterious demand. "The cause of the boom is simple: growing demand in Vietnam and, especially, China," notes the *Economist* with finality.[6] No further explanation is provided—nor required—for readers who are accustomed to accepting a tokenized explanation of exotic Chinese tastes and growing wealth.

Demand reduction campaigns focus on these consumers: How to change their tastes, how to stigmatize their consumption? TRAFFIC, for example, recently published a study interviewing two thousand consumers and more than one hundred retailers across twenty-three Chinese cities about their ivory consumption habits in light of China's 2017 ivory ban.[7] Another study spearheaded by TRAFFIC evaluated eighty-five interventions to reduce demand for endangered wildlife in China to make future recommendations.[8] Studies such as these are quite sophisticated, applying complex statistical analyses to large data sets. They present readers with a wide cross-sectional wealth of findings concerning the modern Chinese consumer of endangered species with the goal of discovering their deeper motivations.

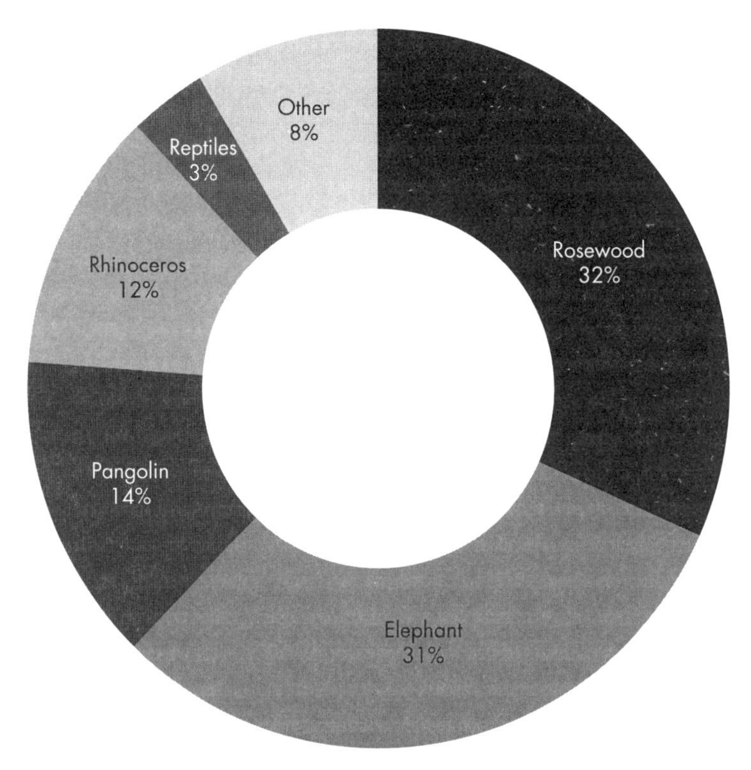

Figure 1.2. Share of type of wildlife among total seizures (aggregated on the basis of standard value), 2009–2013. Reformatted from United Nations Office on Drugs and Crime (UNODC), *World Wildlife Crime Report,* 2020, Figure 4.

There remains a tendency in these studies, however, to reduce diverse rationales for purchasing endangered species to multiple-choice explanations and to distill complex cultural histories into sound bites. In the process, the "crazy rich Asians" explanation is inadvertently reproduced. Large-scale surveys attempt to grasp what drives Chinese consumption, but they fail to grapple with the complex intersection of culture and economy behind the trade. Adding to the difficulty, all of these studies without exception come (however tacitly) with a preordained goal: to mitigate the trade in endangered species. Understanding demand is useful only insofar as it contributes to its future mitigation. Demand reduction campaigns must thus maintain a difficult balance. They have to draw awareness to and stigmatize consumption of endangered species, but they also have to be careful not to stigmatize the broader cultural traditions on which it is based.

An illuminating example is the ad campaign by TRAFFIC and the World Wildlife Fund (WWF) likening the consumption of rhino horn to biting one's nails. Ads feature photographs of rhinoceroses with human hands and toes replacing their horns, with the caption "Rhino horn is made of the same stuff as human nails. Still want some?" The ad campaign is based on scientific evidence that rhino horn is similar in molecular composition to keratin, the primary component of human nails and hair, yet it fails to acknowledge any medical efficacy as reported in experiments conducted in Asia. Drawing from only Western scientific sources with the intent of humiliating a particular group of Asian consumers, the campaign received considerable backlash locally from both government officials and the public, and it was later discontinued.[9]

The medicinal properties of certain endangered species, as this example demonstrates, add another layer of complexity to conservation efforts intended to combat the wildlife trade. Indeed, there can be no real understanding of China's wildlife trade without addressing traditional Chinese medicine. Rhino horn may be carved into a goblet, but it also can be ground into powder and consumed as part of a medley of traditional ingredients to treat rheumatism, fever, and other ailments. Despite its representation in conservation campaigns as mere fingernail powder, the substance has been used as part of traditional Chinese medicine concoctions for three thousand years, and the vast majority of people who lived through the Cultural Revolution will attest to its medicinal powers. Similar medicinal properties are ascribed to tiger bones (as powder or preserved in wine), pangolin scales, and bear bladder, among a long list of other endangered resources.

Not many scientific studies have tested the medicinal properties of these resources, and the results of those we have often differ depending on whether they were conducted inside or outside of Asia. Although Western researchers point to the chemical likeness of rhino horn and keratin, some Asian studies have demonstrated the medicinal effects of rhino horn despite its similarity to keratin. Yet biomedical testing results either in favor of or against traditional Chinese medicine are unlikely to influence adherents. Regardless of their scientifically proven efficacy, an entire generation in China has little doubt of the medicinal value of these substances. "It is unlikely that such deeply entrenched beliefs," ecological economists Hoai Nam Dang Vu and Martin Nielsen conclude, "will be swayed by a one-sided representation of the limited scientific evidence evaluating its efficacy."[10] In other words, as my father-in-law summed it up: "three thousand years—how can it be wrong?"

Traditional Chinese medicine goes far beyond the consumption of rare and endangered species. It is an integral part of everyday life for most people in China, revealing a radical difference from the West in how food and medicine are delineated. Unlike Western consumers, Chinese consumers do not categorize medicine and food separately; rather, everything ingested has potential effects on one's health, contributing to one's internal balance. Traditional Chinese medicine aims to rectify imbalances, and certain substances are particularly helpful in this regard. Bear bile and rhino horn, for example, are particularly useful for dispelling heat, but they do so through different regulatory pathways and thus treat different ailments.[11] They are called for in more severe conditions, whereas other wildlife products can be consumed more frequently. Eating recently slaughtered wildlife, for example, can further intensify the nourishing properties of food. This belief, in part, contributes to the persistence of "wet markets" throughout Asia. The vitality of the wildlife sold in wet markets is an indication of their healthfulness. Wet markets are "wet" precisely because live or freshly slaughtered animals are understood to be more vital to health than those already killed or frozen.[12]

The issue of traditional Chinese medicine thus pertains to not only the illicit wildlife trade but also the issues of wet markets and global health more broadly. The persistence of wet markets in spite of such fierce international opposition demonstrates how ingrained traditional Chinese medicine is, not only in the Chinese health-care system but also in Chinese culture and daily practices. Far from an eccentric tradition that might simply fade away in the face of Western medicine, traditional Chinese medicine comprises the basic fabric of a Chinese approach to health and vitality. The recent inclusion of traditional Chinese medicine in World Health Organization guidelines and as a growing cultural export, as well as its hope for treating COVID-19, confirm that wildlife use and consumption will likely become more prevalent inside and outside of China in the future, not less. This is not to say that wildlife consumption cannot be more safe and sustainable, but Western critiques must be measured and specific; they cannot be conflated with critiques of traditional Chinese medicine as a whole, or they will surely backfire.[13]

The demand reduction campaigns promoted by TRAFFIC, WWF, Conservation International, and so forth must be very sensitive to the fundamental role of traditional Chinese culture in everyday life in China and its diaspora, whether in terms of food, medicine, or home decor. Too often, these demand reduction campaigns fail because their creators do not realize that in attempting to stigmatize wildlife consumption, they appear as though they are stigmatizing traditional Chinese culture more broadly. "Beware of the West attacking traditional Chinese medicine under the name of wildlife protection," reads a prominent headline in Chinese media. "We are the nation that created traditional Chinese medicine which has a tradition of millennia of use," the article observes; "we should not ignore our cultural background and blindly adopt Western values."[14]

Indeed, if not well considered, many wildlife consumption reduction campaigns can appear as an attack on Chinese culture. This certainly applies to rosewood, which plays such a fundamental role in popular legends of imperial China—as I will now show. Ignorant of this iconic history and the role it plays in legitimating and affirming China's contemporary global rise, conservation campaigns aiming to reduce Chinese demand for endangered species are bound to fail.

Ming "cultural bloom": 1400 to 1500

Zhongshan's rosewood mega-mall and other commercial centers selling rosewood furniture are a crucible for a particular type of cultural consumption now in vogue in China. These commercial centers are not simply about showing one has money—although that is certainly done. They are also about reconnecting with an imagined past through one's newfound wealth, as an investment in the future. No one knows this better than those peddling the egregiously priced furniture on the sales floor. Rosewood furniture salespeople in Zhongshan, and throughout China, relentlessly invoke the past. All reluctant buyers on the sales floor, no matter how recently the furniture they are considering was made or how advanced were the engineering techniques used in its manufacture, will hear the story of the ancient craftsmanship that makes rosewood furniture what it is today. Despite now using elaborate machines that produce dozens of engravings in a single day, the rosewood furniture industry still manages to recall the laborious and esteemed cultural work of the dynasties.

This esteemed cultural work began in the Ming Dynasty, considered the golden age of ancient China and one of the earliest embraces of consumer culture in the world. Widely characterized as a period of "boundless extravagance" and "extraordinary prosperity," the late Ming presented the world with "a culturally exquisite bloom."[15] During this time, the sedate certainty of agriculture gave way to the speculative world of commerce as the mercantile economy eclipsed the agrarian.[16] The result was a new—and, one could argue, globally unprecedented—wealth of *things.* Ming material culture ensured that, by 1500, China had "more stuff to think about, or even to think with, than the rest of the world."[17]

Rosewood furniture featured prominently amid this great new wealth of things. As houses became more elaborate, table legs grew, and chairs replaced mats on the floor, furniture began to feature as one of the defining elements of the traditional Chinese home. Adorning foyers and studies, rosewood furniture came to signify the highest levels of cultural sophistication and economic wealth. Characterized by a deep hot wax polish and intricate engravings, classical rosewood furniture originated in Zhongshan and other rural townships throughout southern China in an attempt by timber-rich but otherwise poor areas to achieve maximal value added, given limited resources. From these rural beginnings, the style of furniture gradually gained imperial attention, eventually becoming a quintessential element of Ming, and later Qing, decor.

Social distinctions in early Ming society were quite stark, and rosewood, at least initially, contributed to the demarcation. Sumptuary laws regulated dress and domicile in order to ensure that class membership remained static. This rigid social hierarchy was codified in the Ming Code: gentry and officials at the top, followed by peasants, then artisans, and the much-degraded merchants at the bottom. Only those at the top of the hierarchy could wear certain clothes, eat certain foods, and adorn their homes with rosewood furniture. Thus, initially, rosewood fit neatly within the social hierarchy, demarcating who resided at the top in very clear material terms. Over time, however, rosewood furniture gradually began to engender the reverse. Rosewood and other elite cultural goods became such powerful symbols that, as sumptuary laws eroded, their eventual acquisition by the increasingly wealthy merchant classes began blurring social boundaries.

As Ming consumer culture progressed, the rigidly codified social hierarchy deteriorated. Social relations were increasingly weighed in silver, and it seemed the scales were tipping in favor of the merchants. Despite themselves constituting the lowliest class, merchants gained increasing purchasing power. With unprecedented earnings, merchants now found themselves in a paradoxical position: they were culturally still quite disadvantaged but nonetheless economically powerful in a way that rivaled the gentry. Given such vast merchant wealth, the Ming Dynasty's rigid controls on social mobility faced increasing opposition and eventually dissolution.[18] Sumptuary laws were shamelessly flaunted as the rising merchant class gained possessions formerly reserved for the elite. Wealthy merchants became increasingly indistinguishable from the gentry, previously their superiors.

This inversion of the social order, though not unique to China, did arrive in the country relatively early in world history, and rosewood and other cultural goods played a significant role. Initially reserved as exclusive possessions of the Chinese social elite, rosewood furniture began to trickle down the social ladder. By 1600, there were very few barriers to the downward spread of elite culture, and the homes of even the illiterate nouveau riche often contained elaborate studios furnished with finely crafted rosewood desks and chairs, footrests, and scroll tables.[19] Through their acquisition of rosewood and other elite cultural goods, the merchant class—despite centuries of social oppression—began participating in "high-cultural circles," ultimately "crossing the status barrier between commerce and gentility."[20] The cultural and economic hierarchies of Ming China, formerly distinct, appeared to be "collapsing into each other."[21]

This gradual conflation of cultural and economic power in late Ming society triggered what historian Craig Clunas refers to as the "invention of taste."[22] Taste (*qu*)—that is, the importance not only of wealth but also of the things possessed and the manner of possessing them—became a defining feature of high society in order to preserve the separate hierarchies of culture and economy.[23] The elegant "gentleman" (*yun shi*) was opposed to the vulgar commoner not by his wealth but by the things he owned and his manner of using them.[24] Detailed consumer guidebooks—unique to China at the time—were published instructing the nouveau riche on "gentlemanly" possessions.[25] Such guidebooks distinguished between economic and cultural power—vulgarity (*sú*) on one hand and elegance (*yǎ*) on the other—at the same time that they demonstrated how exactly one might transform the former into the latter. They alluded to, and paradoxically enabled, the social fluidity of late Ming society at the same time they attempted to prevent it.

As one of the most expensive investments for the household, rosewood furniture featured heavily in these texts. Finely crafted rosewood was the ultimate embodiment of elegance over vulgarity. Following the era's privileging of the natural and minimal over the artificial and ornate, noted Ming cultural authority Wen Zhenheng, for example, praised the elegance of hardwood furniture with exposed grain over the vulgarity of similar furnishings painted in gold or overly decorated.[26] The apparent "naturalness" of the wood grain and its durability over the centuries made hardwood implements the foremost candidates for elegant living, as prescribed in Ming guidebooks. To be a good scholar required a study furnished with a rosewood desk, study chair, sitting chair, bench, and footrest to comfortably elevate one's feet while diligently at work.[27] Satisfying what were considered to be the two main criteria of elegance—culture (antique and durable as opposed to modern) and cosmology (natural as opposed to artificial)—rosewood possessions offered a sure route to social status in late Ming society.[28]

In the pursuit of transforming economic success into cultural prestige, rosewood was thus a necessary step. The same is true for ivory, though it was typically displayed less prominently in the household. Ivory, in fact, has an even longer cultural history in China than rosewood, dating back thousands of years, when the material was considered to be more valuable than gold. The tradition of ivory carving was first established in ancient times with mammoth ivory. Trade with Africa subsequently introduced ivory from African elephants, and that soon became the preferred variety. Ivory cut recently after the animal died is most prized, as it is thought to preserve a translucent, ethereal quality of the material. During the Ming and Qing Dynasties, just as with rosewood, ornately carved ivory adorned the homes of the imperial elite, gradually working its way down the social hierarchy (Figure 1.3).[29]

Following the Ming cultural bloom, rosewood experienced a subsequent revival during the early Qing, this time with even more decadent designs. Indeed, the two dominant styles of rosewood furniture—Ming and Qing—are easily distinguished from each other, analogous to the differences between Ionic and Corinthian columns, with the former more minimalist and the latter more indulgent. The classic Ming style is relatively restrained; its curves are usually unadorned and straightforward. The decadent Qing, on the other hand, is much more ornate, covering every available surface in intricate high relief and scrollwork. It should be noted, however, that the decorative elements of Qing Dynasty rosewood are never merely superficial but are full of allegorical and mythological significance.

The term "decorative object" does not quite convey the significance and meaning of these deeply symbolic materials: ivory statues, rhino horn goblets, and rosewood armoires. These items harbor intricate engravings that depict Chinese cultural symbols materialized in the flesh of a natural substance. During the Ming Dynasty, just as rosewood carving was thought to liberate the soul contained in each piece of wood, freeing it within the homes of the imperial elite, the finest carved ivory was considered to have similar powers. Both represented the material convergence of human intention and the cosmic order immortalized in the form of furnishings or implements that would survive for generations to come. These objects united natural forces and artistic proficiency in a material form that remains—despite, or perhaps even more so because of, the Cultural Revolution disruption—the apotheosis of a particularly Chinese style of aesthetic expression.

Cultural Revolution: 1966 to 1976

The imperial heritage of rosewood furniture is paraded without fail, yet its early modern history is often obscured. Specifically, what the rosewood salespeople do not care to mention in their appeals to the cultural history of the craft is its tragic fate during the country's mid-twentieth-century Communist Revolution,

Figure 1.3. An ivory carving from the Qing Dynasty on display at the National Palace Museum, Taipei, Taiwan, November 2015. © George Zhu.

and in particular the decade-long Cultural Revolution. Contemporary vendors, crafters, and investors with whom I have spoken all seemed to brush over this pivotal moment in the history of the industry. Although few rosewood enthusiasts dare to mention it, this drastic yet temporary inversion of the cultural and economic value of rosewood is critical to understanding the wood's present-day resurgence, decades after the chaos of the revolution subsided.

From 1966 to 1976, Mao Zedong led a cultural revolution that upended Chinese society. Mao's goal was to purge all traditional and counterrevolutionary thinking from the country and establish Maoist thought as the sole political and social ideology. Schools and universities were closed; books were burned, assets seized. A generation of bourgeois urban youth—around seventeen million people in total—were "sent down" (*xià xiāng*) to provincial rural farms for reeducation in the wisdom of the humble agrarian proletariat.[30] In the midst of revolutionary fervor, traditional symbols of dynastic power were effaced or destroyed outright. This included Buddhist and Taoist temples, vases and other refined decorative arts, and—without a doubt—any and all rosewood furniture. As an aspirational cultural commodity that signaled a certain form of scholarly success and good breeding, rosewood naturally became a target for Mao's zealous Red Guard, and many of the best pieces on the mainland were destroyed.

Mao's revolutionary class-leveling project was both economic and cultural. The first major redistribution of wealth was spearheaded in the countryside. In one of the most rapid and violent land reforms in history, rural elite were expropriated of their vast landholdings, paraded around in humiliating public rituals, and in some cases summarily executed. Despite the terror, these so-called reforms were one of the most effective, and arguably egalitarian, attempts at redistribution in history. Elite farm size shrank by 90 percent, but that of poorer peasants more than doubled, roughly equalizing the total post-reform landholdings of peasants and rural elite.[31] As a result of this redistribution, the first years of the People's Republic demonstrated an economic rebound in the countryside and a rise in rural living standards as compared with the previous years. Following land redistribution, however, private estates were transformed once more into a system of collective farms, pooled labor, and state procurement, thus abolishing private landholdings entirely. By the end of 1958, 100 percent of rural households were collectivized.[32]

Urban redistribution was more gradual. The expertise of the elite classes was still needed to ensure bureaucratic functioning. Although urban workers were mobilized and the means of production were gradually transformed from private into state and collective property, many elite retained their positions within the industries, now working alongside members of the Communist cadres.

Stripped of their salaries and treated with general suspicion, these elite suffered a severe loss of power. They maintained their positions only to the extent necessary to maintain urban production and bureaucracy. By the 1950s, as the economy throughout China was brought under state control, private markets were replaced by state production, procurement, and rationing and Communist cadres increasingly supplanted the former managerial class.

Economic redistribution in both rural and urban China was drastic, but it was not unique among socialist revolutions of the time. What differentiated the revolution in China was the country's subsequent nationwide campaign to level the cultural field.[33] Redistributing financial assets was not enough, Mao realized, to abolish class distinctions. In 1966, he initiated the Cultural Revolution: the country's most drastic campaign to abolish not only elite wealth but also—perhaps more important—its cultural advantage. By targeting the "four olds" (*sì jiù*) of ideas, culture, custom, and habits, Mao waged an unprecedented strike against class distinction across all of Chinese society. In a complete inversion of reverence for antiquity and refinement, cultural icons—rosewood foremost among them but also jewelry, art, ivory, and other ornamentation—were transformed into shameful relics.

Those I interviewed who lived through the Cultural Revolution recalled the tumult of the time. More than losing their income, families lost their place in society. At any moment, members of Mao's Red Guard could ransack their homes, and rosewood furniture was first on the list of items for expropriation. One interviewee described the gradual pawning of his family's rosewood furniture to make up for salary cuts. When the revolutionaries ransacked his house in search of rosewood and other elite possessions, they were surprised to find no rosewood furniture remaining, as it had all already been pawned. "They could not believe it was all gone," the Shanghai resident recalled. "They searched the house, took what they wanted, and left." In other cases, Red Guards tore up the floorboards in search of hidden items. My father-in-law's house was one such case. He was just a boy at the time, but he recalls his father's pawning of their rosewood furniture before the Red Guard's arrival. "I ate the furniture," his father responded when asked what had happened to it all, their family's enviable collection of rosewood furniture amassed over generations. This was his clever and perhaps irreverent way of telling his son, my father-in-law, he had pawned it all to buy food.

At the height of the Cultural Revolution, no one dared to be caught harboring rosewood furnishings or other bourgeois indulgences. To avoid confiscation, families sometimes painted or altered their rosewood possessions to make them unrecognizable. To this day, it is possible to find rosewood tables and chairs once engraved with faces of ancient Chinese figures now defaced and disfigured in an attempt to erase their elite associations.[34] Rosewood furniture that was not destroyed was often redistributed to farmers in the countryside. Antique tables and chairs now worth a fortune would furnish the workstations of rural carpenters or—if wood supplies were low—serve as fuel to heat their houses. Consequently, much of the best and most historically significant rosewood now sits in museums in Taipei or Hong Kong, where exiled elite hurriedly exported their collections after the rise of the Communist Party.

During the Mao era, the Ming Dynasty's "invention of taste" rapidly deteriorated, as "all people, no matter what level of cultural capital they possessed, were forced to have the same taste."[35] Desiring anything other than Communist propaganda made one the subject of public humiliation and a principal target of the party.[36] The atmosphere of attack and suspicion that permeated the Cultural Revolution—as neighbors, workers, and students were all encouraged to wage spontaneous, violent assaults on the former elite—ensured the drastic devaluation of any cultural possessions that might be construed as counterrevolutionary. Reverence for traditional icons was replaced by reverence for products of the revolution: pictures of Mao, red badges, army attire, anything symbolizing allegiance to the movement.

After Mao's death in 1976, the Cultural Revolution ended. Those sent down to the countryside returned to the cities, and the great purge of traditional icons subsided. Exactly how to move forward in the post-Mao era remained a hot topic of debate for the upper echelons of China's Communist Party. In 1978, Deng Xiaoping became the next paramount leader of China, succeeding Mao with a platform based primarily on "restoring order" (*bōluànfǎnzhèng*) after the Cultural Revolution chaos. Although restoring social order was Deng's initial—and less polemical—political objective, he is most famous in the West for what he accomplished next. Following the easing of social tensions, Deng and the party began considering how to move forward economically. Although Deng is hailed in the West as a type of laissez-faire Chinese prophet—sowing the seeds of free market capitalism, which at the time was expected to inevitably blossom into liberal governance and Western-style democracy—he was in fact a highly pragmatic leader with little regard for ideological doctrines, liberal or otherwise. His early dealings with capitalism were pure experimentation—experimentation that happened to be highly successful, given the vast flows of capital released by the global neoliberal order.

Deng began his experimentation first in marginal areas that were not vital to the state-led economy. These zones became known as "special economic zones," with Shenzhen in southern China being the first.[37] As these zones demonstrated remarkable success, the model was reproduced on a larger scale. In executing this approach, Deng followed a "no debate" policy. Since most in the Communist Party did not agree theoretically or ideologically on the best path forward, they had to experiment in practice.[38] Unlike Russia's catastrophic "shock therapy," China deployed markets not as an end in themselves but as a means to achieve political legitimacy and further solidify political stability and control. Under Deng's model, market logic was first and foremost subordinated to the logic of the state. Yet, though the goal was political stability and control, the result happened to be the most laudable economic growth streak in history and, along with it, a complete revitalization of Chinese traditional culture.

cultural boom: 2000 to present

By the late 1980s, Deng and a newly liberalized Communist Party opened strategic regions in China to foreign investment and economic modernization, eventually overseeing the largest expansion of the middle class in history. Although capitalist dynamics within China remained subordinated to party doctrine, the booming economy in the country's special economic zones enabled many individuals to gain financial success almost overnight, further igniting the desire to seek refuge in class distinction and cultural elitism.[39] From the country's first moves toward reform and opening in 1978 until its accession to the World Trade Organization in 2001, China's per capita gross domestic product (GDP) increased by nearly an order of magnitude, from US$156 to US$1,053. Over the next twenty years, the country's per capita GDP increased by another order of magnitude, reaching US$10,262 in 2019, on par with Brazil and South Africa and far above India.[40] In terms of total GDP on a purchasing power parity basis, China is the largest economy in the world.

With steady GDP growth and now more billionaires than the United States, China is a major source of luxury consumption, yet the trends do not always map onto those of the West. Although urban China provides booming retail centers for Western luxury brands, reviving staid traditionalists such as Burberry and Coach, a large portion of Chinese spending is dedicated to the purchase of categorically Chinese products, with traditional Chinese objects representing an increasing share. Classical icons formerly scorned by the Communist Party now feature prominently within the Chinese imaginary as a source of national pride.

For those who came of age during the Cultural Revolution, it has become difficult to reconcile Maoist ideology—the promise of a just and equal society rising out of the flames of a decadent classical China—with China's new economic liberalism. Effectively, Deng's economic revolution swept aside Maoist ideals, leaving a cultural vacuum. Without the imperatives of Maoist doctrine, and unwilling to fully embrace Western consumer culture, many Chinese are now looking to buy back—or even "steal" back, in some cases—their lost cultural heritage. For the Chinese middle classes who grew up under Mao and got rich under Deng, endangered resources such as rosewood furniture, ivory statues, and rhino horn goblets represent more than just Veblenian conspicuous consumption; they represent an opportunity to reconnect to a rich cultural heritage denied under the asceticism of the Cultural Revolution. That is, endangered cultural resources offer a way to regain a lost sense of Chinese identity while still abiding by the capitalist mandate to consume.

This is not to say that buying rosewood and other cultural resources has nothing to do with conspicuous consumption. The scene at Zhongshan's furniture exposition demonstrated very clearly that indeed, people were there to be seen. I recall in particular, among the tumult of buying and selling, two men dressed nearly identically, circling and finally boisterously easing into the chairs of a Qing Dynasty–style living room set with theatrical groans. Each occupied a chair, with a coffee table between them and a hardwood "sofa" on the far side—all ornately decorated with phoenixes and dragons, polished to an ostentatious sheen.

"How much for the set?" one of the men asked the sales rep in a booming voice.

"250,000 RMB" (approximately US$40,000), she replied nonchalantly.

The two buyers looked at each other and shared wide grins. "Good. I'll take the whole set," the other announced, continuing to insist that "this is the one. I need this delivered today." The sales rep apologized with a tight, embarrassed smile, trying to explain that those were just floor models from the factory, examples of a new season of wares, and would not be deliverable for months. This sparked a round of indignant and loud argument from the two men; almost immediately, a small audience gathered around them on the showroom floor. My husband and I watched in quiet disbelief. Yet it soon became clear that the buyers did not mind the scene they had created; indeed, they were reveling in it. They started throwing around numbers and figures: they need five sets—no, six—but all by the end of the day. A large part of this row was intended for the audience now thronging the vending stall as the two men communicated in a comically exaggerated fashion not only their wealth and importance but also their sophistication and taste.

The scene certainly retained elements of conspicuous consumption as commonly understood. It demonstrated, as with Ming Dynasty mercantilism, the blurry line between commerce and culture, having money and having taste, identifying with one's cultural heritage and simply showing one is rich. Indeed, both are achieved in the purchase and display of rosewood furniture, and this is precisely what makes this particular consumer revival so powerful. Yet the consumer revival is only half the story. While China's nouveau riche travel across the country to Zhongshan in search of elaborate dining sets to display in their homes (and perhaps the opportunity to perform their newfound wealth), so too do those looking to buy the wood solely for its ability to appreciate in value and who need not revel in their purchases so conspicuously. The cultural appeal of rosewood has not been lost on investors looking to hedge inflation and diversify their portfolios. Lasting for generations, endangered rosewood finished in the form of furniture provides an unparalleled investment opportunity in contemporary China. "With such rare wood," one salesman assured me, "the price is sure to grow." Because of its scarcity, longevity, and potential for appreciation, rosewood furniture is now purchased more as a financial investment than a cultural memento.

Rampant financial speculation in China—not just in rosewood but across the board—has been difficult to control since the early 2000s. In other Asian countries, such as Thailand, Indonesia, Malaysia, and South Korea, this began earlier, triggering the financial crisis at the end of the 1990s. In China, volatile capital flows started in the early 2000s and continue to this day. The financial term for these flows is "hot money"—speculative capital investments that are channeled very quickly from one country to another to take advantage of varying interest and exchange rates. China is a notorious recipient of hot money, which has been estimated to reach the trillions. From 2003 to 2008, an estimated US$1.75 trillion in hot money accumulated in Chinese markets as a consequence of favorable exchange rates and in anticipation of future appreciation of the RMB.[41] National monetary policy has been implemented to restrict these sudden inflows and outflows of capital, to varying effect. In 2015, for example, after technical adjustments that inadvertently devalued the RMB were misread by speculators as a sign of weakening currency, the country faced rapid capital flight. The Chinese government responded by tightening regulations for Chinese groups investing overseas, requiring approval for cross-border deals worth more than US$10 billion and for state-owned enterprises investing more than US$1 billion in foreign real estate.[42]

Hot money flows, compounded by the speed of economic growth in China, have placed increasing pressure on traditional investment avenues (stocks and real estate), making nontraditional investment avenues such as art and endangered species an increasingly appealing alternative. The Chinese government has been at pains to combat speculative bubbles in its stock and real estate markets, with municipalities introducing draconian policies limiting residential purchases to one per family in order to avoid speculation. Interestingly, this has inadvertently resulted in a spike in divorce rates among wealthy urban residents seeking a second property. My cousin-in-law in Shanghai was among the statistics when he and his wife "divorced" in order to purchase a second residence and quickly remarried after the sale closed. To curb such practices, municipal governments have implemented countermeasures limiting homeownership for the recently divorced.[43] When such speculative sleights of hand run out, residents are forced to look elsewhere for investment. Rosewood furniture has become one of the alternatives.

The speculative demand for rosewood in China is well documented. Those within the industry constantly point to the investment value of the wood as the primary driver of the market boom after 2005. Indeed, rosewood has become its own type of stock exchange. Not a single person working in the industry who I interviewed, nor any who were interviewed by Chinese media outlets I surveyed on the topic, failed to mention speculation as a (if not the) primary driver of the boom. "Compared with the vagaries of real estate and the stock market," one market analyst observed, "the price of rosewood furniture has been growing a lot more steadily."[44] "They use it as a bank," an interviewee summarized the dynamic to me in terms I would understand, knocking on a wood table nearby to demonstrate the durability of the investment.

Although in many ways the speculation benefits the industry, importers and manufacturers have mixed feelings about it because it adds a volatility that can be difficult to manage. Rosewood furniture factories have been known to go bankrupt because the order price they accepted months prior ended up being too low to buy even the timber required to fill the order. By 2007, timber supplies were so low that one factory owner found himself offering gold bars in exchange for logs of the most endangered woods.[45] The rosewood market became "a playground for investors rather than a regulated market for the collectors or homeowners who admire it."[46] By 2008 the bubble burst, bankrupting many furniture retailers and factories. Quickly rebounding, however, the market returned to what was considered a "healthy" condition—that is, catering more to regular buyers than speculators. Yet by 2010 the market was again in a "price heat" and again "largely due to speculative trading" as opposed to personal purchases.[47] "Currently there are few attractive channels available to hot money," explained one market analyst in 2011; "given the doldrums in the housing market, the depressed stock market and the risky futures market, rosewood furniture has quite naturally become the target of many investors allured by its appreciation prospects."[48]

By 2013, China's rapid economic growth triggered renewed speculative trading and a new round of rosewood price spikes—this time the highest yet. Prices for the most expensive rosewood species had increased by more than 500 percent per year since 2005 and annual market sales surpassed US$25 billion in 2014.[49] Many of the family members I interviewed whose rosewood possessions were confiscated during the revolution bought new rosewood furnishings to replace them. But as of late, prices have become too extreme. "For the best wood, now you must be very rich," one interviewee observed, rationalizing his purchase of a lesser-valued rosewood. Consumers are now turning to cheaper varieties sourced from new locations. Meeting this demand for new rosewood varieties has required a dramatic acceleration of rosewood logs imported from across the tropics. China's rosewood imports averaged as much as 180 percent growth per year, with much of this growth coming from Africa (Figure 1.4). Rosewood imports from Africa experienced an overall increase of 700 percent from 2010 to 2015.[50]

When I began my fieldwork at the end of 2014, the market experienced a modest downturn, but it remained inflated well beyond 2005 levels. "There is value, but no market," a number of interviewees at one of the country's largest rosewood markets observed.[51] At the time of my interviews in 2015, the price of the wood was still quite high, but few were selling. As traders held on to their stock, confidence that the market would rebound was palpable. "These markets go through cycles," one interviewee noted nonchalantly over a glass of tea, surrounded by piles of logs that he was not yet prepared to sell, given current market conditions. Beginning near the end of 2016, rosewood prices did indeed begin to rebound. When I visited furniture manufacturers in the spring of 2018, production continued with the hope that prices—still quite high, although low compared with the 2013 peak—were only going to get higher.

It is no coincidence that the rosewood market has, to a large extent, mirrored the market for Chinese art and antiques—rising in the early 2000s, bubbling by 2007, experiencing a significant downturn after the global financial crisis, but also recovering fairly rapidly thereafter. Both markets have faced intense speculation because of an oversaturation of other, more conventional forms of investment in China. The Chinese taxi-driver-turned-billionaire who spent a record-setting US$36 million on an eight-centimeter "chicken cup" once looted from the imperial Summer Palace, only to drink from it in front of a stunned crowd, is just one of a deluge of Chinese investors who survived the chaos of the Cultural Revolution and are now looking for outlets for their newfound wealth.[52]

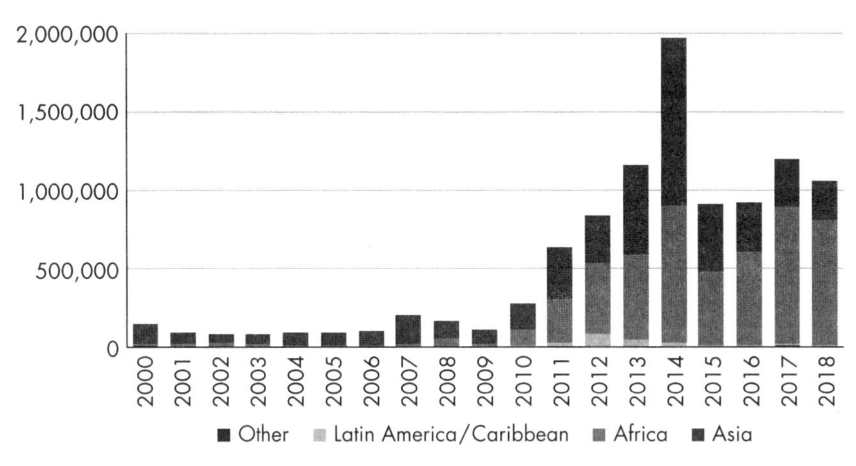

Figure 1.4. China's rosewood imports (roundwood equivalents in cubic meters), 2000–2018. *Data source:* China Customs, compiled and analyzed by the Forest Trends Association.

More recently, breaking this record but with less of a spectacle, a modest Chinese brush-washer dish sold to an anonymous phone buyer for nearly US$38 million.[53] In an even more extreme example, setting a world record for Asian art across the board, a Chinese vase initially fired in the imperial kilns of the Qing Dynasty and subsequently "discovered" during a routine house clearance in a London suburb, sold for more than US$69 million despite its meager initial asking price of US$800,000.[54] Such record-breaking investments create a climate ripe for speculation. "The market for historical Chinese art is so frenzied," wrote one reporter, "that even seemingly mundane pieces of Chinese art can electrify the scene at auction homes."[55]

A similar dynamic characterizes Chinese contemporary art, although with crucial differences. The case of Chinese contemporary art demonstrates how in many ways an object's "Chineseness" matters more than its desirability when it comes to financial speculation. This is because, unlike rosewood furniture and traditional art, Chinese contemporary art is not universally esteemed within the country. In fact, at the time of its initial boom in the early 2000s, the genre had very little cultural appeal in China. Most investors regarded the works—with their ridiculous mixing of Mao and Marilyn Monroe or recaricaturizing of Communist cadres with violent laughter—as an absurd anomaly, barely worth the canvas on which they were painted. Yet despite their palpable disdain for the genre, China's financial elite swallowed their "cultural prejudices in order to embrace Chinese contemporary art as an alternative mode of investment"—solely because of its appreciation prospects.[56] These investors, mostly Chinese, purchased the works not because of their strong cultural value but simply because all things distinctly Chinese are booming. Whether they personally identified with the art was less important than the fact that Chinese people in the future might.

As a consequence of this speculation, by 2011 China became the largest contemporary art market in the world, accounting for nearly half of global sales. Today, approximately one-third of the top-selling five hundred contemporary artists worldwide are Chinese.[57] Chinese buyers are "redrawing the maps," with contemporary Chinese artists dominating the list of top five sellers and Chinese auction houses rivaling those of the West in sales turnover.[58] Western art critics value contemporary Chinese art because it plays with classical Chinese icons in sometimes irreverent ways that draw from Western artistic traditions. But Chinese investors push the genre to new financial heights, investing in it simply because it is Chinese. It is on account of its speculative potential, much more than cultural value, that Chinese contemporary art has gained such value.

This, too, is the case with rosewood. Not long ago, an antique rosewood corner-leg table sold for nearly US$4 million at auction, more than fifty times its initial estimate of just more than US$80,000.[59] Soon after that, I came across a Ming Dynasty–era rosewood stool selling on eBay with an asking bid of US$3 million. Through either antique items or newly manufactured furnishings, the rosewood market represents a ready buying opportunity for financiers looking to store value, hedge inflation, and diversify their portfolios. With each distorted by the other, cultural and economic values collapse within the furniture itself, deconstructing and reconfiguring ancient symbols to serve the speculative economy. As one of China's most notable contemporary artists, Ai Weiwei, has metaphorically demonstrated, the chiseled frame of the rosewood market has warped under the pressures of a late capitalist economy (Figure 1.5). It is unclear where culture ends and economics begins. The cultural history of China is being reinvented, repurposed, and reconfigured within the context of the country's global economic integration. More than tradition, rosewood now signifies an "invented tradition"—or, rather, a *re*invented tradition.[60] Conjuring fragmented memories of an imagined antiquity torqued by global financial flows and the impetus to accumulate, the rosewood market is not old but rather new indeed.

Valued for both its cultural prestige and financial appreciation, rosewood offers a unique perspective on China's booming cultural economies. Chinese merchants during the Ming Dynasty used the growing social prestige of rosewood to convert their burgeoning economic capital into a tool for ascending the social hierarchy in the fifteenth and sixteenth centuries. In contrast, as the cultural value of rosewood plummeted during the country's mid-twentieth-century

Figure 1.5. A reconfigured Qing Dynasty rosewood table, *Table with Two Legs* by contemporary Chinese artist Ai Weiwei, demonstrates the distortion of traditional values in contemporary China. San Francisco's Asian Art Museum exhibit *28 Chinese*, 2015. © Annah Lake Zhu.

Cultural Revolution, the wood's economic value plunged as well, becoming almost worthless in China's collectivist economy. After the country's reform and opening, China's new consumer class has engaged in a renewed round of distinction making, buying back rosewood stolen from them during the revolution. Wealth is converted into cultural distinction through the purchase of rosewood and other elite cultural goods by China's rising nouveau riche. Jumping on this growing consumer demand, aspiring financiers are purchasing these items not because of their cultural value but rather because of their prospects for appreciation. These financiers convert cultural distinction back into wealth through speculative investment.

The cultural history of rosewood mirrors China's historical transformation more generally. Amid the country's rapid global integration, rosewood has in its own small way come to represent what it means to be Chinese in history and Chinese in the world. The market's reinvention of traditional values in light of late capitalist dynamics is characteristic not only of rosewood but also of the contemporary Chinese experience writ large. The current demand for rosewood and other endangered species cannot be understood outside of this experience. Just as tropical deforestation symbolizes one of the greatest planetary threats within the conservation imaginary, the cultural history of rosewood and its contemporary revitalization symbolizes a continued cultural eminence and national cohesion within the Chinese imaginary. In a revival of styles that date back to the Ming and Qing Dynasties, rosewood furniture has come to exemplify China's contemporary cultural identity; but it has also come to exemplify how consumption and investment, cultural history and hot money financial flows, have become welded together in the heat of the present cultural boom. This fusion, demonstrated so clearly in the case of rosewood, is the defining feature of nearly all aspects of Chinese consumerism in the twenty-first century.

2

hot money capitalism

Northeastern Madagascar is home to one particular variety of rosewood that fuels China's cultural boom. This region is a type of island within an island, isolated from the rest of the country with no paved roads in or out, but at the same time connected to the global economy via lucrative markets for rosewood and vanilla, the region's other major export. Both of these commodities have highly volatile prices that fluctuate by orders of magnitude from year to year, month to month, sometimes even day to day. Rural residents feel these drastic price shifts. In this region of Madagascar, China's dematerialized hot money financial flows swiftly materialize throughout the rural countryside in high-denomination notes of Malagasy money. The money is wielded in large suitcases, hidden under straw mattresses, buried in rice fields, or spent in egregious sprees.

Upon arriving in northeastern Madagascar, by either an hour-long flight or a turbulent two-day bush taxi ride from the capital city, one of the first things most foreigners notice is the overabundance of "exotic" flora and fauna. Chameleons lounge in the bushes. Hotels keep lemurs as pets. Enormous breadfruit and jackfruit bend down toward the forest floor as if begging to be eaten. At times, mangoes litter the ground like fallen leaves. Combined with the beauty of the landscape—the turquoise hues of waves lapping against pristine white sand beaches, the vine-tangled hills that rise over the few low buildings of the towns—daily life in the region appears to take on a dreamlike quality.

Things in the sleepy coastal towns get much stranger, however, during rosewood and vanilla booms. While walking along a dirt road, for example, one may see a chameleon plastered in money. At the fruit stand down the street, a farmer who typically makes a little less than a dollar per day might buy a crate of mangoes and turn to smashing them on the ground one at a time. Emptying his pockets, he might then deliver a fistful of cash to the stand owner—more than covering the damage—and stroll away, content. These seasons of fevered expenditure are not marked on any calendar nor strictly pegged to any natural cycle. Rather, they depend upon fluctuating global demands. Fueled by China's booming markets in Zhongshan and across the country, Madagascar's rosewood economy channels large and sudden influxes of cash into remote forest villages that otherwise see very little. Vanilla—the other major source of foreign exchange in the region—does the same. Mirroring rosewood market dynamics, vanilla follows even more extreme oscillations. A kilogram of cured beans can range from less than US$5 to more than US$500, earning a rural farmer either US$250 for his annual crop or more than US$25,000. In either case, his earnings will typically be paid to him in a lump sum, on his straw bed, in his *ravenala* hut, with the nearest bank days away. If not stolen by traveling thieves, the money is likely to burn a hole in his pocket—or even his mind.

It is within this context—baseline poverty punctuated by extreme abundance—that one might find money glued to a chameleon or the purchase of mangoes just for smashing. Such profligate acts of spending are also referred to as "hot money" (*vola mafana*) in Malagasy vernacular, showing clear parallels with but also sharp differences from the global form. Hot money is a notorious feature of the social landscape of northeastern Madagascar during commodity booms. As my friends in Madagascar recounted these fantastical spending stories to me, I listened in disbelief. Emaciated rosewood loggers emerge from the forest in tatters but with more money than ever before sealed in plastic bottles tied around their necks. Impoverished vanilla farmers become Malagasy "vanillionaires" overnight. Greeted by such wealth, they do strange things. One farmer, I was told, bought fancy shoes only to go around town stepping on feet and blaming it on "the vanilla." Another traveled to the town festival in order to buy up all the rings at a ring-toss stall and collectively throw them in the opposite direction of the pyramid of bottles that should have been their target. As the crowd looked on in amazement, he slapped his hands together and joyously proclaimed, "That's how you play with money!"

No one knows how better to play with money than the rosewood and vanilla workers of northeastern Madagascar. Yet, in addition to play, this newfound wealth occasionally leads to more tragic displays. In a more sobering tale, one farmer allegedly boiled up all his earnings in a type of money-soup and, after ceremoniously consuming it, was said to be found dead the next day. Rumors and stories abound, not only about how people spend their money but also about the products themselves: vanilla not as a delicate flavoring but as the vital ingredient in dynamite or tires; rosewood not as luxury timber but as a cover for the traffic in human bones. Often unsubstantiated, such allegations nonetheless take on a life of their own, inundating the region—like the punctuated wealth brought on by the export commodities themselves—with mysterious and fantastical undertones of alternating abundance and dearth.

Hot money in Madagascar shows a very different side from hot money globally—at least at first sight. How can one explain these extravagant bouts of spending, seemingly tantamount to throwing money away? Or the extravagant stories that gravely misconstrue foreign demands for Malagasy products in terms of dynamite or bones? Against the already enchanted surroundings, it may appear that such flagrant and exuberant acts of spending are irrational or tinged with a kind of magical thinking, but the reality is more mundane. As bizarre as Madagascar's hot money spending might seem, it is based on a logic that is all too familiar and, upon further scrutiny, does not depart that drastically from the logic of hot money in the financial sense.

The financial term "hot money," as noted in Chapter 1, refers to speculative capital flows that are channeled very quickly from one country to another to take advantage of varying interest and exchange rates. China's booming rosewood market is without a doubt the product of these hot money financial flows, as investors find novel cultural channels for reaping short-term financial gains. Vanilla, too, chronically suffers from short-term speculative bursts, similar to hot money finance. Reports of looming crop shortages or new global demands send investors into a purchasing frenzy in which they pounce on potential returns. Hot money, as understood in economics, is thus the product of intensive financial speculation characterized by large and sudden flows of capital following a short-term investment horizon. It is the foremost source of the extreme instability that defines capitalist development in an era of global finance—what some have aptly termed "casino capitalism."[1] Rather than a system of rational economic progress, the new global economy more closely resembles a high-stakes game of fortunes won and lost. Governments such as China's are at pains to control these unwieldy, opportunistic financial flows.

Hot money in Madagascar is not unlike hot money globally. In fact, it provides another side to the story. Beyond the centers of global financial speculation, hot money dynamics also penetrate its margins. As elite financiers roll the dice on speculative investments channeling capital into China's rosewood industry or Madagascar's vanilla markets, Malagasy loggers and vanillionaires navigate the resulting booms and busts. Their bouts of extravagant spending—not at all illogical—offer a sobering reflection on our global financial system of "hot money capitalism," the effects of which now reach the farthest corners of the world.

Of course, not all of rural Madagascar's rosewood and vanilla earnings are spent in hot money sprees. Only during times of extreme market boom do vanilla and rosewood workers engage in these acts of extreme spending. (And even then, only some of them do.) Chapter 3 tells of the more mundane investments—land, roofs, cows, computers—in which money is subject to the productive capitalist dynamics one might expect. But just like hot money globally, which circulates as an opportunistic distortion of more productive capitalist circuits, hot money in Madagascar crops up during moments of spectacular excess. Both are extreme deviations from the conventional capitalist norm, yet in their deviation they reveal the driving force of capitalism. They respond to the paramount requirement of the constant movement of money—sometimes quite literally "throwing it afresh into circulation," as Karl Marx phrased it long ago—but in unconventional ways that can have far-reaching consequences, as this and the previous chapter show.[2]

This chapter uses the boom-and-bust dynamics of northeastern Madagascar to illustrate China's rosewood market madness from a different angle. The hagglers, window-shoppers, lowballers, buyers, sellers, and speculators of Zhongshan—much like the Westerners who critique all this consumerism—have very little understanding of where the majestic wood that will furnish their homes for generations actually comes from. More than this, they rarely consider how these rapid transactions amounting to billions of RMB might materialize in a place where the virtual economy remains next to nonexistent. Illuminating this other side of hot money flows, this chapter uses rosewood—and also vanilla—to analyze how the speculative world of global finance manifests in northeastern Madagascar, starting with the vanilla boom in the early 2000s, then moving on to the rosewood boom less than a decade later, and concluding in an even more extreme vanilla boom, which has recently turned to bust.

The chapter shows that hot money in Madagascar abides by a logic similar to that of hot money globally. Money in a capitalist system must always be in motion. Surpluses must be channeled. With few productive outlets for capital, Malagasy people are finding their own channels. Those who happen to arrive in northeastern Madagascar during the seemingly exotic hot money season are not encountering illogical or exotic behavior but rather a unique cultural inflection of the all too familiar oscillations of global capitalism. Any exoticism we might find is little more than a distorted reflection of the exotic roller-coaster ride that is our global financial modernity.

magic in the market

"Fantastical and magical reactions" to the global economy, as anthropologist Michael Taussig phrases it, have long been documented across the globe.[3] From spirit possessions in factory labor to witchcraft in modern politics, magic appears to pervade responses to the market and modernity at the farthest corners of the world.[4] The connection may at first seem puzzling. Why would magic and fantasy—in some senses the supposed antithesis of modernity—be a common response? Yet magical responses to modernity and the market are not a product of esoteric traditions stubbornly cropping up in the face of their modern antitheses. Rather, they reflect and often resist the magicalities inherent in modern forms. They "show the extent to which modernity—itself always an imaginary construction of the present in terms of a mythic past—has its own magicalities, its own enchantments."[5]

Anthropology is rich with ethnographies of the magical and seemingly irrational blossoming in the face of modernity. These "exotic" responses are not backward or absurd but reflect the exotic fictions—the commodity fiction, phantom objectivity, myth of modernity, and so forth—hidden in everyday capitalist modernity. Rumors of the boundless power of witches, for example, mirror the opportunities for boundless abundance offered by modern politics.[6] Likewise, the fetishization of evil in the form of the devil reveals an analogous commodity fetishism hidden in the market model.[7] Studying fictions created at the periphery of global capitalism, in other words, makes it easier to decipher those fictions that compose its core.

Madagascar's export economies, however, do not follow the development trajectory implicit in some of these classic accounts. The seminal cases of "proletarianization" in South American mining operations described by June Nash and Michael Taussig, for example, consider the mines to be "a synecdoche for the modern age of industrialization."[8] The history of the mines "encompasses the rise of an international expansion of capitalism that exported capital and machinery from the metropolitan centers to the farthest corners of the world."[9] The mines embody the quintessential "precapitalist" to "capitalist" transition. In export economies surrounding cash crops, similar processes of capitalist transition and class formation have also been observed.[10]

In contrast to these accounts, Madagascar's vanilla and rosewood export economies have instigated very little capitalist "development" or class formation in the traditional sense of the terms. These export economies are part of a wider network of informal export operations that persist and even proliferate within the new global economy.[11] They are characterized by precarity and indeterminacy rather than hopes of modernization or progress.[12] Despite being linked with the cosmopolitan luxury demands of a speculating global elite, rosewood loggers and vanilla growers supplement subsistence-based lifestyles through fragmented networks of manual production that have changed little in more than a century. They provide a synecdoche not for the modern age of industrialization, but for the connection between pre-industrial extraction and agriculture on one hand and post-industrial financial speculation on the other.

Under conditions such as these, the cultural phenomenon of hot money, as well as the stories and rumors that circulate and document it, becomes increasingly prevalent. Mining booms provide classic examples, including gold rushes in the Brazilian Amazon, ad hoc dollar economies in the diamond mines of Angola, and coltan booms in the Democratic Republic of the Congo.[13] In Madagascar, hot money is a cultural trademark of the more volatile export economies. The practice has been documented in the country's northern sapphire mines, the clove and vanilla regions of eastern Madagascar, and the illicit "business" (*biznesy*) of urban Tamatave (Toamasina).[14]

Academic explanations of hot money spending have historically reduced the practice to either a passive adaptation rooted in a "culture of poverty" at one extreme or a romantic tendency rooted in a lifestyle of the present at the other.[15] Eschewing these two extremes, a more promising interpretation acknowledges that hot money is neither passive nor particularly romantic but rather is an active response to an imposed marginality.[16] In this reading, extravagant bouts of spending conjure a sense of autonomy. They perform a type of "semiotic 'magic,'" inverting the marginality of the spender in a "momentary realization of fantasy" and control.[17] Like mushroom foragers conjuring liberation amid precarity, profligate spenders attain momentary autonomy, a cathartic release from the uncertainty of the new global economy.[18] Other explanations point to the social capital profligate spenders often acquire.[19] But, in its most extreme manifestations, hot money spending may result in social stigma and a severe loss of control and autonomy (the recently paid farmer incapacitated by the sight of his earnings, for example, or the logger who has spent all his money in a single spree, ostracized and abandoned the next day).

Of course, the diversity and complexity of hot money in Madagascar and elsewhere make it impossible to reduce the practice to a single cause or explanation. Yet there is a particular logic exercised in these creative and occasionally unsettling displays, a logic that suggests an alternative interpretation. Like other fantastical reactions to the global economy, hot money is a creative cultural inflection that both reflects and resists the volatile economic dynamics Malagasy producers experience every day. With all its ups and downs, its spectacular alternations of abundance and dearth, hot money spending demonstrates not the backwardness of magical thinking in an area left behind by capital but an acute cultural expression of global finance capitalism.

from vanilla to rosewood and back again

Northeastern Madagascar, as noted, is a rather insular place (Figure 2.1).[20] Rural villages lack electricity and running water and are often days away from the nearest financial institution. Roads within the region are well paved, but those connecting the region to other parts of the country are unpaved—in the rainy season nonexistent—and take days to traverse, sometimes at the risk of one's life.[21] Outside the cities, life is in many ways what one might imagine when dreaming up an African pastoral landscape driven by subsistence rice cultivation, including the grind of planting and harvest; the bumpy dirt roads connecting scattered villages that only pedestrians, motorbikes, and a few specially rigged four-by-fours dare to traverse; and the makeshift stands by the side of the road, where villagers leisurely gather, selling dried rice and beans, fresh fruit, avocados, and other odds and ends. Yet, alongside this rugged and relatively globally disconnected pastoralism, northeastern Madagascar maintains global connections—and occasional influxes of large sums of money—through its rosewood and vanilla export economies.

Despite the financial speculation that surrounds them, remarkably little has changed in both rosewood and vanilla production since these export economies were established on the island. Rosewood used to grow throughout Madagascar's eastern tropical forests, but it was logged to near extinction in the southeastern and central forests during the colonial period and is now relegated almost exclusively to the northeast. Because of its colonial associations, rosewood is considered "a tree with a hard history" and has become "coded to locals as trees of exploitation."[22] Rosewood harvesting methods, as discussed further in Chapter 3, have changed little since the colonial period, with axes used to fell and prepare the logs and as many as sixty men needed to haul them out of the forest by rope. The work remains largely unmechanized because of the difficulty of transporting heavy machinery into the forest and the machinery's generally high cost and low availability. As was true centuries ago, human labor continues to be much cheaper and more abundant. Despite the extreme difficulty of the work involved, loggers make what they largely consider a respectable daily wage (about US$3–$5 per day, or 10–20,000 ariary). Mid-level traders profit a

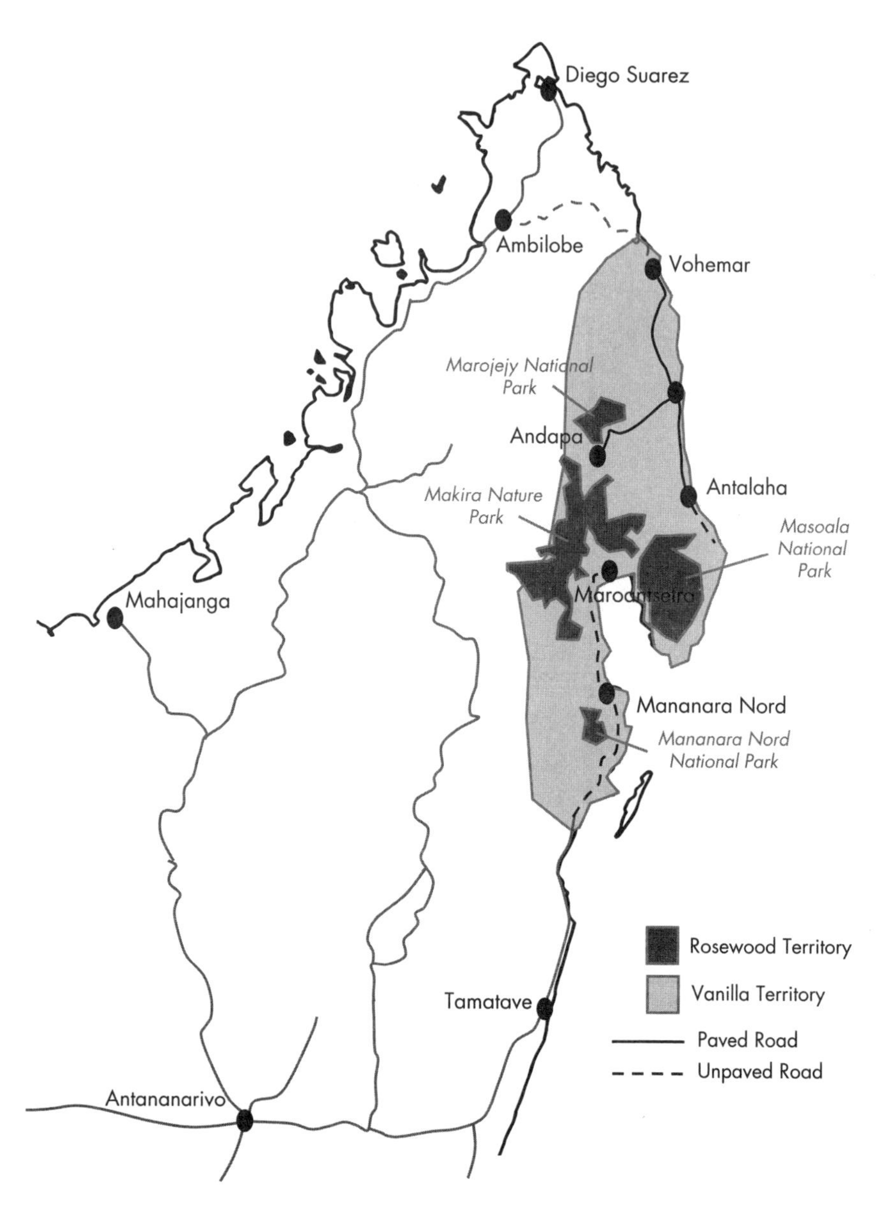

Figure 2.1. Northeastern Madagascar's rosewood and vanilla territories. Note that the only roads connecting this region to the rest of the island are unpaved roads (dashed lines), which can barely be called roads at all. © Annah Lake Zhu.

great deal more, whereas the exporters, in contrast to both, have become some of the richest people on the island.

Vanilla has a similar supply chain on the production side and similar colonial associations. Native to Mexico, the vanilla cultivated in Madagascar (*Vanilla planifolia*) now generates nearly 80 percent of the global supply. Thick and waxy, the plant is an orchid that grows like a vine. It is punctuated by occasional clusters of white flowers that, if pollinated, will each yield a long, green vanilla bean (Figure 2.2). With no natural pollinator, vanilla in Madagascar must be pollinated by hand, making it one of the most labor-intensive cash crops. Introduced during the colonial period, vanilla cultivation continues on small family plots throughout the countryside. For the most part lacking economies of scale, growers still use basic techniques introduced by the colonial regime, planting vine by vine with basic shovels and pollinating flower by flower with no more than a splinter delicately bending the anther. The typical vanilla farmer subsists on less than a dollar per day outside of boom times, though mid-level vanilla traders (as with rosewood traders) profit a great deal more.

Those who export rosewood typically also export vanilla. Now very wealthy, these exporters, referred to as the "bosses" (*les patrons*), consist of a small handful of families, most of whom have lived in the region for nearly a century. Many of them are in fact of mixed Chinese-Malagasy origins, established in Madagascar after a wave of Chinese immigration beginning in the early twentieth century. This wave of Chinese immigration reached other African countries as well. After the fall of the Qing Dynasty in 1912, at the dawn of China's "century of humiliation," during which the country would be plagued by European and Japanese invasions, many people living in China sought work and a new life overseas. Especially after the British and French abolition of slavery in the mid-nineteenth century, thousands of Chinese immigrants were brought to Madagascar (and other colonies) as indentured servants for large-scale agriculture and infrastructure projects.[23] Although some workers returned to China, many stayed in Madagascar. A large number traveled from the central highlands, where they worked on railroad construction, to the northeast, where the land was fertile and there was money to be made in cash crops.

These were the "old Chinese": largely Cantonese-speaking people from southern China who, unlike the Mandarin-speaking "new Chinese" of this century, came to Madagascar to stay. They live in the region to this day, with Malagasy passports and large families, and have come to dominate both the rosewood and vanilla export industries. First, they exported vanilla, a relatively easier cash crop with which to begin an export operation. As demand for rosewood from China picked up in the early 2000s, some of these families leveraged

Figure 2.2. Green vanilla beans on the vine. © George Zhu.

their historical Chinese connections and began exporting rosewood as well. Over the past two decades, many of the bosses of the region have switched from rosewood to vanilla and back again as the markets waxed and waned. Together, these bosses employ thousands of men and women to perform the tedious tasks of heating, drying, curing, and sorting their vanilla, and thousands of loggers and traders when the rosewood market is at its peak. They own major real estate in and around town—palatial houses, shops, and boutiques. Much like the subtle fragrance of drying vanilla beans that saturates the region in late summer, the exporters increasingly organize life in the region according to their own field of gravity.

Although, on the supply side, the harvesting and production of rosewood and vanilla have changed little since colonization, the demand for these products and their producer prices have transformed dramatically during the past two decades. Vanilla was first. Beginning in the mid-1990s, at the insistence of international financial institutions such as the World Bank and the International Monetary Fund, the Malagasy government ended a policy of price setting that had been in place for more than twenty years. This market deregulation was followed in 2000 by one of the largest cyclones in Madagascar's history, which devastated prime cropland and sent vanilla prices into uncharted territory. By 2003, the price of vanilla reached more than one hundred times its historical levels as a result of rampant speculation in the face of limited supply. The vanilla boom of the early 2000s "brought eastern Madagascar into the compressed space-time of globalization."[24] Before this, intensive speculation was unheard of: the price was low, but it was predictably low and growing at a steady 2–3 percent per year.[25] Rural residents rarely saw such returns as would permit a culture of savings. They simply had no need for banks and no impetus to consider their earnings beyond the six-month horizon over which their money would surely be spent. But along with the millennial price spike came a dizzying influx of money. Still echoing its history of colonial oppression, vanilla began to also present unprecedented opportunities for capturing the wealth and abundance of the new global economy.

Amid the hype, residents with no experience in growing vanilla began joining the booming market. They gathered cuttings and watched their vines grow for the three requisite years before they might bear fruit, tending them eagerly. But just as these aspiring farmers harvested their first crop, the market collapsed. The price of vanilla fell sharply three years after its peak. Weathering the crash, farmers sold their motorbikes and mattresses and went back to sleeping on straw beds. The crop came to symbolize, in yet another dimension, the exploitation associated with the outside world.

But all was not lost. As the price of vanilla reached its lowest point in 2007, exporters in the northeast began to realize a new opportunity: Chinese demand for rosewood. Having already experienced the single-greatest enrichment of the region in decades, vanilla exporters took no haste in entering the rosewood market after the vanilla market crashed. "The decline of the vanilla industry, a critical mainstay of regional elites," a World Bank report noted in 2010, "opened the door to the establishment of rosewood market activity on the back of the vanilla market."[26] Given the exporters' preexisting ties to China, the transition proved surprisingly smooth. After a coup d'état in 2009 left the country with a semilegitimate "transition regime," a type of lawlessness swept over the country and the rosewood market opened for export (see Chapter 3). Vanilla exporters hastily switched to rosewood.

As the price of vanilla began its tragic descent, the price of rosewood began climbing in a near mirror image. The region's vanilla boom-and-bust was mirrored in rosewood logging, which, by 2009, saw a similar price spike in the region and similar demonstrations of hot money spending. Villagers throughout the region abandoned their rice fields and entered the forest by the thousands in search of rosewood. By 2012, the number of rosewood exporters mushroomed from just thirteen after the coup to more than one hundred.[27] After the installment of the new government in 2013 and strict government prohibitions in 2014, however, exports sharply declined and the price of rosewood plummeted, shutting many new exporters out of the market.

But yet again, in a volleying of profits from market to market, all was not lost. As the logging boom began to subside in 2014, the price of vanilla began another rapid ascent, this time dwarfing the price spike in the early 2000s. Rampant speculation and hoarding in the aftermath of another cyclone in 2017 drove the global price of vanilla to more than US$600 per kilogram—more than double its price during the millennial boom and surpassing the price by weight for silver. Some are quick to suggest that laundered rosewood money also contributed to vanilla undersupply and speculation, driving up the price.[28] Awash in cash from their rosewood exports, the region's elite allegedly turned to buying up vanilla to launder their earnings. Indeed, growers and traders all over the region confirm that there is a deep connection between rosewood and vanilla market dynamics, as clearly expressed in their inverse ebb and flow.[29] Now, as the vanilla boom is beginning to subside—with export prices nose-diving from more than US$600 per kilogram to $300 and falling—exporters and residents of the northeast are considering again turning to rosewood.

late capitalism in rural Madagascar

Representing some of the only ways to obtain cash in the region, rosewood and vanilla have brought the volatility of global financial speculation to a place with few formal mechanisms for coping with its drastic oscillation. Although price instability is unavoidable in a scarcity-based capitalist market, neoliberal reforms and the financialization of the global economy associated with late capitalism greatly exacerbate the phenomenon. From vanilla to rosewood and back again, northeastern Madagascar has experienced ceaseless cycles of boom-and-bust, now coming from either end of the world in rapid succession. Indeed, since 2000 all of the region's major cash crops—including coffee, cloves, pepper, and cocoa—have experienced unprecedented price volatility that contrasts with the relative stability of earlier years. The past two decades have made irrefutably clear that extreme booms and busts are the norm, not the exception.

"Late capitalism"—a term popularized by critical theorist Fredric Jameson, following Marxian economist Ernest Mandel—describes the most recent era of capitalism, ongoing since the 1970s but still very resonant today.[30] In both academic and popular discourse, late capitalism continues to capture the dematerialized, consumer-driven economy characterized by instant communication, ephemeral desire, perpetual disposability, and outsourced labor. For Jameson, late capitalism has resulted in "a kind of cyberspace in which money capital has reached its ultimate dematerialization, as messages that pass instantaneously from one nodal point to another, across the former globe, the former material world."[31] It represents the "purest" stage of capitalism yet.[32] Through the financialization of the economy, "money becomes in a second sense and to a second degree abstract. . . . Like the butterfly stirring within the chrysalis, it separates itself from that concrete breeding ground and prepares to take flight."[33] The abstraction of money under late capitalism thus becomes complete—dematerialized to the greatest extent possible.

But in northeastern Madagascar, money remains far from fiction. Here, a dematerialized virtual cyberspace comes in brutal contact with the still very material world. Butting into this stubborn materiality, global capital must be translated back through a disjointed conversion that sends paper money spitting out into the material landscape as if randomly by machine. Profits throughout the rural countryside are collected and wielded in large suitcases. The money might burn. It will likely be stolen. It lies in an uncomfortable lump under one's mattress, a constant reminder of its painful materiality.

With little access to the virtual economy that dictates the demand for its products, rural communities living near areas where rosewood and vanilla grow find it difficult to cope with the materiality of money. The banking system is largely inaccessible and underused. Madagascar's Central Bank estimates that the equivalent of US$340–$450 million is held by residents of the northeast outside of the banks. Corroborating this trend, those I interviewed estimated that banks are used by only about 5 percent of vanilla farmers, 1 percent of rosewood loggers, and 10 percent of mid-level traders (in both vanilla and rosewood). Cooperatives exist across the countryside, sharing techniques and social support, but they rarely pool their money or provide insurance, given the liability associated with holding large sums of cash. Without savings and investment, one sees "signs of money everywhere, but few signs of wealth."[34] This is equally true for rosewood loggers living in rural areas. Even urban rosewood traders, who can more easily access banks, are discouraged from using them because of the illegality of the trade.

Life at the nexus between the virtual and the material becomes almost surreal. Farmers who manage to grow any vanilla at all during these times can make previously unfathomable amounts of money. So too can anyone with a log or two of rosewood. "Ironically," it has been observed, "the influx of money in a culture where money was not traditionally a primary objective, and where the concept of expendable cash has never been an option, has caused problems of its own."[35] Loggers and rural farmers are targets of theft, and the export markets begin to resemble the drug trade. Vanilla farmers sleep with machetes in their fields to prevent theft off the vine. Shipping containers of vanilla are welded shut each night and torched open the next morning. Villages are occasionally raided by armed gangs, and roadblocks are set up to ambush passing traders. To counter the chaos, the government has at various times sent in the military. As the price of vanilla began to rise again during my fieldwork in 2015, similar transgressions resurfaced and police stationed themselves at the side of the road. Cars were stopped, and friends coming home late from the discotheque were frisked for vanilla or cash.

The problem is not only local. Even if banks and credit were available to free vanilla growers and rosewood loggers from the materiality of their earnings, insecurity in the national financial system continues to plague Madagascar's economy. In December 2001, for example, just as the price of vanilla made its first spectacular ascent, the country nearly fell into a devastating civil war. As one news outlet described the situation, "mercenary gangs fought with demonstrators, commerce was blockaded, and the banking system shut down while thousands of businesses failed."[36]

In 2016, as vanilla prices ascended once more, the country experienced rapid inflation, making savings unappealing. In the midst of the inflation, the American spice giant McCormick & Company anticipated price hikes and allegedly withdrew Malagasy money from banks by the crateful in order to buy vanilla—and possibly to prevent others from doing so as well.[37] In the northeast, the banks ran dry. "We couldn't buy vanilla for three days," a smaller company owner noted, "until the government printed more money and sent it up here. It was crazy."[38] This is a common occurrence at harvesttime during boom years. Even the exporters lack faith in the system, sending much of their money overseas, not only to evade national authorities for illegal earnings but also to sidestep the unstable domestic economy.

All this is to say that money in rural Madagascar is painfully material in its comings and goings. Its volatility is based on the far-off demands of a speculative market that deems vanilla the "no longer boring" flavor of the future and exaggerates the scarcity of supply or, in the case of rosewood, rebrands classical Chinese furniture from the dynasties into a new investment opportunity with great appreciation prospects.[39] These speculative fluctuations have drastic consequences for the communities of northeastern Madagascar. Global vanilla prices can double in a single day, triggering a mad rush to wire transfer money to the exporters and buy while supplies last.[40] Buyers must wait days for more cash to arrive from the capital. Similarly, with the rosewood market, looming global trade prohibitions can send timber speculators into another mad rush to secure the last of the supplies. In the export markets of northeastern Madagascar, global space-time meets the snail's pace of island country life, reminding us that those experiencing the symptoms of late capitalism are often the most removed from its accoutrements.

export culture

Volatile economies shape cultural life. This is one of Fredric Jameson's fundamental insights, although any observer of mass media, fast fashion, and the like might come to this conclusion. Referring to the "cultural logic" of late capitalism—ephemeral consumer desires, fragmented visual reproductions, and schizophrenic signifiers with no referents—Jameson highlights one side of this dynamic. He shows how cultural life under advanced capitalism begins to embody the inherent volatility of the new global economy. But there is another late capitalist cultural logic of which one might speak. This is not the cultural logic of late capitalism generally but rather that of its still very material margins. It is the cultural logic of a place where advanced capitalism meets its opposite: where paper money is transported in wheelbarrows and automated teller machines run dry for days.

The cultural logic of late capitalism in northeastern Madagascar is what I, following Malagasy people living in the region, call "export culture."[41] Export culture might also be summarized as *ny toe-tsaina amin'ny asa fanondranana vokatra,* "the mentality of working with export commodities." In referring to export culture, Malagasy people highlight the link between volatile market dynamics and moments of hot money spending. Building on this view, I use the concept of export culture to refer not only to extreme spending practices but also, more broadly, to the sense of disjuncture and mystery that stems from a profound reliance on trade in faraway spaces. Export culture arises from the tenuous connection between an advanced global economy and its underdeveloped margins. In northeastern Madagascar, export culture manifests in sporadic displays of waste and excess, as well as elaborate fictions that reframe the rosewood and vanilla markets in a fabulous utopian space, revealing in the process a deep ambivalence to the outside world.

Madagascar's export culture does not belong exclusively to the market booms of the past two decades. As a long-standing feature of Malagasy society, acts of profligate spending have been documented since the establishment of global trade routes in the seventeenth century and well into the colonial period.[42] Stories of earlier spending practices now inspire residents who are experiencing the most recent volatility. The vanilla farmer's feet-stepping provocation discussed above, for example, follows a tradition established by early coffee farmers who benefited from a particularly good crop.[43] But what was before an occasional outlier—a combination of exceptionally high yields and modest price increases—has increasingly become tuned to the erratic rhythms of the global market.

Hot money spending is a primary example of export culture. First, during the millennial vanilla boom, it was the older vanilla farmers who engaged in the practice. Elderly farmers bought fancy bicycles they had not yet learned to ride, only to walk them up and down the street on display.[44] Years later, the rosewood boom has shifted the demographic of hot money spenders from older farmers to younger loggers. Alongside this demographic shift has come a shift in spending. Echoing the older vanilla growers, young loggers bought motorbikes they were afraid to ride, only to park them outside their homes for neighbors to stop in and have a closer look. These loggers might then sell their motorbikes—or new phones or computers, whatever they had purchased—months later for half the price.

In addition to consumer items, entertainment is another likely channel for hot money spending. With the transition from older farmers to younger loggers, entertainment venues have also shifted. In the logging camps, makeshift casinos and discotheques were quickly erected to cater to younger logging crowds. Similar impromptu establishments sprouted up in logging villages throughout the region. In coastal cities, recently paid loggers invited their friends to a night at the bar to proudly display the array of empty bottles their group had consumed. One bar in particular—Discothèque Bois de Rose—got its name, I am told, from the young rosewood workers who would arrive, ask how much it would cost to rent the bar for a night, pay the stated fee, and stay just long enough to down a shot of the bar's finest liquor.

Along with the thrill of hot money spending, there is a sort of madness that occasionally stirs with the acquisition of such unseemly sums. As with the phenomenon of hot money generally, the potential for such madness has been documented for quite some time. Early coffee farmers with exceptionally fruitful harvests were often susceptible. When being paid, these farmers would occasionally interrupt: "Stop, stop! It is too much money."[45] Again, this story resembles one that I heard of a contemporary vanilla farmer being paid in his home by a traveling trader. As the trader laid down each bill in counting, the farmer grew increasingly agitated. With only half his money paid, he suddenly exclaimed that the money was too much and politely asked the trader to return with the remainder a different day. An even more extreme story is that of a vanilla farmer who sold 500 kilograms of vanilla and promptly went mad.

From consumer binges to near insanity, stories of hot money spending reframe the export markets so that all members of the community may participate, even if only by rumor and gossip. "It makes people laugh because it's so crazy," a resident observed of these stories, "[but also] in another side it makes people sad." As a kind of dark comedy, these stories gain a certain autonomy, becoming memorialized in local folklore. As they circulate throughout the region, they are shortened and generalized, acting almost as fables. People revisit them in times of market lull and draw on them for creative inspiration as the markets inevitably bring yet another influx of cash to the region and yet another round of hot money hysteria.

mystery markets

Despite more than a century of trade in rosewood and vanilla, these markets retain deep-seated mysteries on both the supply and demand ends. Rosewood loggers and vanilla growers know little about how their harvests are ultimately used, even though they understand with great clarity "the extent to which their fates [lie] in the hands of powerful, even mysterious, foreign others."[46] On the flip side, rosewood and vanilla consumers in China or the West have symmetrically little understanding of these luxury goods for which they may or may not, depending on the moment's frivolity, be willing to pay so much money.

Take, first, the mystery of rosewood. As noted, the rosewood logging boom has made Mandarin "the language of money" in the northeast, and many young Malagasy people are eager to learn it. Yet, though most residents know that rosewood is headed to China, neither loggers nor low-level traders know why this tree specifically is of such spectacular value. When I asked loggers about how the wood is used and why it costs so much, their answers varied widely, from "to build houses" or "for medicine" to "it's just a famous tree!" As one *New York Times* article (somewhat misleadingly) remarks, "Francel, like others who carry axes into the mountains, finds it curious that rosewood is so valued. Other trees yield food—papaya, coconut, jackfruit."[47] Contrary to the implications of this quote, rosewood workers indeed appreciate the utility and aesthetics of wood. Malagasy residents often prefer, however, the light brown palisander over the dark violet rosewood (although rosewood has gained in popularity since the logging boom—see Figure 2.3). There is certainly a great deal of confusion as to why their preferred wood fetches, at times, less than one-tenth of rosewood's price.

Vanilla shares a similar mystery. Vanilla growers and driers have little idea how consumers use the spice, and they certainly do not use it themselves. Each farmer I asked shook his head with a smile, gesturing that he is not quite sure where his vanilla is going nor for what it will be used, except that it goes "outside" (*ivelany*) to foreigners like me. Some offered vague possibilities, such as that it would be added to coffee or rum. Later, I relayed my surprise to one of my friends. How could these workers whose sole income relies on vanilla not know what its use is?

"If one person knows, then I think everyone would know," my friend replied.

"So you don't think any of them know?" I asked.

"No, I don't think any of them know, only the bosses," he declared.

I pondered this for a moment. "Wait, do you know?"

Another moment passed before he admitted with a resounding laugh that even he did not know and then asked me to tell him. I explained that vanilla is used for taste and smell—cooking and perfumes—but I realized that this was a poor explanation. As I searched for a better explanation, I was reminded of a passage in Tim Ecott's book on vanilla concerning a similar exchange with a group of workers during peak prices in the early 2000s.[48] These workers did not believe that vanilla could earn so much for only its taste or smell. They guessed instead that it was used to make what they considered much more consequential products—dynamite or tires. My friend expressed similar reservations about my explanation, as though he had heard it before but did not quite believe it. Surely there was some greater purpose for this mysterious bean that employed, as he asserted, up to 70 percent of the population in certain cities during the drying season.

Frustrated by my inability to convey the ultimate use of vanilla, I further confused things by introducing the concept of fake vanilla, and another disjointed

Figure 2.3. Wooden ornaments (*sary sokatra*) sold at a local market. The dark red ornaments are rosewood, which has gained in popularity since the logging boom in 2009. © Annah Lake Zhu.

exchange followed. I told my friend that it is actually quite easy and cheap to synthesize vanilla flavor with almost exactly the same effect as the natural variety.

"Oh," he responded, becoming quite interested. "Can *you* make it?"

"Can *I* make it?" I laughed, almost appalled. "Of course not. I haven't the slightest idea how to make it!"

And here arises the other side of the mystery. Consumers, of course, have symmetrically little understanding of the distant source of these highly coveted luxury items. Few of those who are enticed by its delicate flavor know that vanilla is the fruit of a Malagasy farmer's early-morning labor. They likely do not even know that it was once a fruit at all (myself included, at least until one of my Malagasy coworkers, surprised and amused by my ignorance, introduced me to my kitchen by way of the forest). And, across the globe in China, the rising urban elite who adorn their houses with rosewood have equally little knowledge of the distant source of this immensely esteemed wood. Indeed, as I traveled to Shanghai after my trip to Madagascar, I found a number of rosewood-owning residents surprised to see a photo of the wood's bright red splinters lying at the hands of a Malagasy woodworker. Even rosewood importers who I interviewed in Shanghai had very little idea of how they ultimately obtained their Malagasy rosewood.

What is, for the consumers, an indulgence of the most inconsequential kind is, for the producers, a lifework. Given this discrepancy, it becomes much easier to understand why farmers and loggers might circulate other tales regarding the commodities that compose the bulk of their livelihoods. Far from ignorant misconceptions, these stories—such as that vanilla is the vital ingredient in pivotal industrial products, as discussed previously, or that rosewood exporters make money from bones rather than from rosewood, as discussed in the section that follows—are ways to understand one's daily toils in a more meaningful light. They rewrite the export markets in more favorable terms, obscuring the somber fact that one's lifework is, in another world, but a niche luxury good tossed about by the turbulent seas of a late capitalist market.

bones and moneymaking magic

Many residents of northeastern Madagascar, I came to learn while living in the region, did not believe that some of the most elite rosewood exporters have actually made their fortunes from the timber. Instead, these exporters were the focus of elaborate rumors involving bone theft and moneymaking machines.

One rosewood exporter in particular, unanimously agreed to be the richest in the region, features prominently in them. Surprisingly, in the rumors retold to me, the source of this exporter's vast wealth was attributed not to rosewood but to either the illicit trafficking of human bones or a Chinese-bought moneymaking machine (*machine manao vola*). Even after the exporter declared, under pressure, the source of his earnings and admitted on national television to being a "pioneer" in the rosewood industry, people still spoke of bones and money machines. They gossiped that his machine had been confiscated, and they wondered whether the bone theft would continue. Yes, apparently this multimillionaire had worked with rosewood, but that was of only minor interest to most in the region.

Using bone theft or a moneymaking machine to account for this exporter's wealth, I now understand, actually makes a great deal of sense—much more sense, in fact, than any other explanations of rosewood or vanilla possibly could to most Malagasy people. How better to explain the region's intermittent influx of cash, for example, than by a machine? Residents throughout the northeast know that rosewood is bought along the coast by Chinese ships filled with Malagasy currency. That this money might be obtained in advance through a bank in China was a surprise to everyone I interviewed. A Chinese-made money machine that prints Malagasy money is a much more compelling explanation, especially given that the region's most prominent exporter has been rumored to have acquired one for himself.

The trade in bones is another compelling explanation for such a lucrative businessman. Bones are by far the most important material substance in Malagasy culture, and rumors of their theft have been long-standing throughout the island of Madagascar, even though there is little evidence of such an export market. In a ritual exhumation called the *famadihana*, bones provide the material vehicle through which relations with the ancestors (*razana*) are conceptualized and performed.[49] With such great cultural importance, what better explanation for the wealth of the region's richest individual? Indeed, talk of bone theft around town has heightened since the rosewood and vanilla booms in the northeast, and a number of exporters have been accused of being involved in it.

By attributing the wealth of prominent exporters to bones and a machine rather than rosewood, residents of the northeast are, once again, participating in a mysterious global economy, even if only through rumor and gossip. Residents suspect that rosewood exporters are doing something devious and deceitful. But rather than explaining this deception as the theft of a natural resource, which makes a great deal of sense to conservationists throughout the world, Malagasy people prefer to understand it in their own terms. Their stories critique what is perceived as the ill-gotten gains of the local elite, but they do so through a distinctly Malagasy register that does not rely on foreign notions of endangered species or natural wealth. Such rumors transform the ultimate source of power and wealth from rosewood or vanilla (both culturally inconsequential products used for who knows what on the other side of the world) to bones (the most profound symbol in Malagasy culture) or a machine (a much more sensible encapsulation of otherness). This revision makes the happenings around town more familiar and engaging. It brings the exotic into the realm of the everyday. The Malagasy themselves—rather than the foreign demands for their products—are placed at the center of the story.

subverting the power of money

Northeastern Madagascar's export culture—hot money spending and fabulous stories that surround the market and its major players—reflects the extreme market dynamics that penetrate the region. Hot money spending mimics the volatility of the region's unpredictable export markets. Similarly, fabulous stories that recast rosewood as bones or vanilla as dynamite reflect a parallel transformation within the global economy. Just as global markets transform luxury indulgences at capitalist centers into the principal livelihoods of those living at the margins, in an analogous conversion, fabulous stories reinvent these minor luxury commodities. Through fictitious transformation, these stories bridge the global (dis)connect that matches superfluous demands at one end of the world with vital economies at the other.

But more than innocent reflection, export culture also retains a subversive potential. Both hot money and fabulous stories reinvent the established order, removing—albeit ever so slightly—the power of the market over daily existence. By bringing global market dynamics into the realm of the everyday, they perform a kind of minor resistance. Hot money subverts the power of money by being spent freely: it transforms what might otherwise be seen as the sacred lifeblood of the productive apparatus—money—into a fleeting binge or amusing spree. Acts of frivolous spending and the stories that depict them demonstrate the power and hold that money has over the region, but also the opposite. Through their lively and sometimes tragic show, creative displays of excess rid one of money as quickly as it has arrived: "they consume money itself," eating money (*mihina vola*)—sometimes literally—rather than seeing it eaten by someone else.[50] Treating money so cavalierly—eating it, throwing it away, or spending it freely—reduces its power. It subverts an economic order based on the strategic accumulation of money. It also softens the madness that occasionally stirs when one confronts, in moments of extreme market boom, the sheer magnitude of one's earnings.

Stories concerning the mysteries of northeastern Madagascar's export markets perform a similar subversion, creating a world insulated from the realities of the established order. Through these stories, rosewood is transformed from a market for an unremarkable wood into a market for human bones. Demand for rosewood thus becomes demand for the most consequential cultural material on the island. Through a parallel fiction, vanilla is transformed into the vital essence of an explosive powder, making it a major player in the story of industrialization.

The subversion performed through acts of hot money spending and fabulous storytelling is in many ways symbolic. Through narratives that disrupt and reorganize meaning, fabulous stories reinvent trivial niche markets into powerful symbols of Malagasy culture (e.g., bones) or global industrialization (e.g., dynamite or tires). Hot money spending performs a similar type of subversive symbolism. By spending money freely, consumers symbolically empty money of its function, thus evacuating its power. In the process, the identity of money "as a symbol of economic dependence is inverted and defeated."[51] Taken together, fabulous stories and hot money spending generate a subversive energy that resonates in undertones throughout the region, symbolically challenging the power of money and the market. This can be seen all too clearly, for example, in northeastern Madagascar's vanilla farmer who, in boiling up his earnings and eating it as soup, symbolically subverted the power of money in a poignant performance with tragic results.

But the subversion is also material. That vanilla farmer—ridding himself so quickly of so many bills—was also physically disposing of an extravagant sum of money that might otherwise have gotten the best of him (although it ended up doing so anyway). This farmer and other hot money spenders across the region free themselves of the materiality of money. They creatively navigate an unpredictable economy that sporadically materializes in suitcases of Malagasy currency wielded throughout the rural countryside. Rather than keeping the money under their mattress or lugging it to the bank, Malagasy people spend it freely. Hot money spending thus embraces the absurdity of an overwhelming abundance of a resource that is otherwise quite scarce. In the process, money loses its power. It becomes a material substance just like any other: a substance that comes and goes, a substance to eat or throw away.

Madagascar's rosewood and vanilla boom profligacy recasts the excesses of capitalism in new light. Sums reaching the trillions flood Chinese markets, exploiting favorable interest and exchange rates, driving up stock and real estate prices, and eventually reinventing traditional rosewood furniture into a novel investment avenue with surprising consequences for rural Malagasy residents. Likewise, flavor and fragrance companies in the United States and Europe speculate over the year's crop, and rural Malagasy residents again brace themselves for the resulting boom. As problematic as this free-flowing capital is, threatening financial crises across the globe for economies that cannot manage the liquidity, it would be far worse if the money were to stop flowing altogether. Hot money is destructive and destabilizing, but the far greater threat is the rigid crystallization of capital in place—the rigor mortis of money.

Hot money spending in Madagascar and hot money speculation globally maintain strong parallels. As capitalist dynamics require, both keep money in motion. But they do so in "unproductive" ways that take the logic of capitalism to very different, yet equally absurd, conclusions. Hot money in its financial manifestation creates surplus value and, in a sense, economic growth, but only by exploiting what might be considered loopholes in the global financial system. It creates profit without actually being productive; it is opportunistic and exploitative, as with all capitalist dynamics, but not like the mutually beneficial "invisible hand" of classical economics. No greater societal good is achieved via the pursuit of individual self-interest. Hot money in Madagascar is also unproductive in the classical economics sense, yet, in contrast to financial hot money, it directly embraces its lack of productivity. Madagascar's hot money spending transforms the sacred lifeblood of the capitalist apparatus into a mere amusement or toy. It provides almost a parody of global capitalism, abiding by the need for money in motion but with none of the intentionality, as if blind to the sole purpose of the game.

Hot money in both its Malagasy and global financial manifestations thus provides absurd caricatures of capitalism classically conceived; each represents two sides of the same story, opposing faces of the same volatile dynamics. Yet, whereas the latter learns all the rules of the game only to exploit loopholes, to win the game by sleight of hand, the former refuses to play, undermining the whole pursuit by failing to abide by the singular logic of putting money to work making more money. Rather than exploiting loopholes, hot money in Madagascar cuts the Gordian knot; it lights the game on fire. By spending all their money in a single spree, Malagasy farmers and loggers reveal in its starkest relief the duality of money in a capitalist economy. Money is both highly sought after and carelessly thrown away; it is the most important thing, and yet it must always remain in motion. This duality is not unique to money in Madagascar but is the hallmark of money circulating globally. Money is, at one extreme, the most important thing in fulfilling one's livelihood, but also, at the other extreme, something one *cannot* hold onto. Vast excesses must be channeled and absorbed, redirected and repurposed. As fundamental to meeting one's basic needs as money is and as dearly as one might want to hold on to it, money in a capitalist system must necessarily be tossed about as if it were hot.

The duality of hot money spending in Madagascar—extravagant spending that trivializes money, compounded by the unshakable awareness that money is far from trivial—reveals the global logic of money. It is only as this logic plays out across the material landscape that its dynamics are made abundantly clear. Not at all primitive or backward, northeastern Madagascar's export culture reflects the material experience of an otherwise dematerialized global economy. Rural Malagasy farmers plastering money onto chameleons are not demonstrating extreme behavior or misguided backlash; they are simply channeling capitalist excesses in a place ill-equipped to absorb such windfalls. With few outlets for speculative financial flows, rural residents have little choice but to embrace their own channels. As hot money flows from China's booming timber markets to the forests of Madagascar, rosewood loggers and vanillionaires find themselves navigating the most turbulent seas of today's speculative economy from afar.

This dynamic makes us question what is at the heart of the contemporary capitalist system, to which there currently exists no foreseeable alternative—real or imagined. Is it the freeing of resources to gravitate toward outlets of highest return, maximizing efficiency and enabling growth? Or is it money in motion, opportunistic and unproductive, for better or for worse? Hot money in Madagascar and hot money around the world suggest the latter, albeit from very different angles. They provide caricatures of the system of rational economic progress we call capitalism. Juxtaposed, they raise the question of whether one should continue playing the game to absurdity or light it on fire.

3

taking back the forest

Clouds of black smoke loomed over the capital of Madagascar in a sudden outburst of political turmoil on January 26, 2009, later to be deemed Black Monday.[1] It was the first coup d'état in more than three decades for the island nation—although controversy, scandal, and political deadlock had surrounded nearly all of the country's presidential elections during that period. Black Monday was particularly destabilizing. The day was named after its tripartite "accumulation of darkness": the rising black smoke from scattered arsons, the dark moral depravity of the looting crowds, and the deep obscurity through which a small group of ringleaders had allegedly masterminded the scene.[2] The violence that erupted that day inaugurated a two-month period of acute unrest across the country, culminating in the military's seizure of the Presidential Palace in the capital city and the forced installation of Andry Rajoelina—the opposition leader—as the head of the new government, later deemed the High Transitional Authority.

Rosewood both fueled and profited from the chaos following Black Monday. The upside of the political unrest (*tombontsoa ny grève*), residents of the northeast told me, was the opening of the rosewood trade (*misokatra andramena*). These residents insisted not only that rosewood logging benefited from political turmoil in the capital but also that money generated through the trade partially financed the unrest. Reporters investigating the trade make a deeper allegation—that rosewood money has been financing both the country's presidential elections and the political unrest that inevitably follows since at least 2000.[3] Indeed, this lucrative export economy has had consequences far beyond hot money spending. Whereas Chapter 2 focused on the extreme economic oscillations of the rosewood trade in Madagascar, with sudden inflows of money into rural areas and few outlets for savings and investment, this chapter explores the political dimensions, along with the details of everyday life and work in the forest. Although rosewood has been a part of the Malagasy economy since before the country's colonization in the late nineteenth century, only since the collapse of the Malagasy government via a military-backed coup d'état in 2009 has there been such an outbreak of logging, dramatically changing the daily lives of the residents in northeastern Madagascar as well as the overall political geography of the country.

In the wake of the Black Monday protests and to no one's surprise, the unrest in northeastern Madagascar gravitated toward rosewood. After ransacking and burning shops owned by the faltering president, Marc Ravalomanana, rioters made their way to local offices of the Ministry of Environment and Forests, where hundreds of previously confiscated rosewood logs lay piled next to the building. Trucks owned by the region's elite reportedly shipped in rioters to storm the offices and shipped out the confiscated logs to their own private estates. Within minutes, hundreds of logs worth millions of dollars disappeared. The ministry offices were looted and files and equipment destroyed in a demonstration that lasted all night. Amid the chaos, two sport utility vehicles belonging to the Madagascar National Parks system were smashed and overturned. Outside the cities, residents stormed the national parks in search of rosewood. The road leading from one of the major parks in the region was reportedly streaked red with the remains of the deeply hued logs dragged from the forest. During the next two months, thousands of containers of rosewood logs were exported from the region, making a few select exporters multimillionaires.[4]

This was the beginning of what would later come to be known as "the time of the rosewood" (*lera ny bois de rose*): the largest unrestricted export of endangered rosewood logs the country had yet to experience, lasting on and off for the next five years and generating revenues of more than US$1 billion for a small handful of exporters with connections to Chinese markets. Leveraging their record profits, some of these rosewood exporters have since managed to secure seats in the Parliament of Madagascar following the country's return to electoral politics in 2013. As a new power bloc within the central regime, they were alleged to have spread "rebel money" (*vola miodina*), buying votes to impeach the president and destabilizing the government from within. Presumably, their hope was to trigger a new period of instability and a new rosewood free-for-all. As confiscated rosewood logs piled up across the region, these exporters-turned-parliamentarians positioned themselves as saviors of the trade. They urged the government to reopen the trade and led strikes when their voices were not heard. To this day, they continue to feed on the growing discontent of their constituencies in the northeast, who—disillusioned by the exclusionary policies of the national parks (see Chapter 4)—now sit on the dormant munition of as much as US$2 billion in cached logs already cut and awaiting export if only the government would allow it.[5]

As is true in even the most advanced democracies, political triumphs are often best understood through the eyes of their constituents. Tens of thousands of workers all along the rosewood trail who have in their own eyes made it big from the trade—even if by simply acquiring a proper bed on which to sleep—propel the politics of rosewood beyond the heights it might reach through only a handful of elite exporters. The rosewood exporters of northeastern Madagascar know this and cultivate a Trumpian appeal. Bolstered by self-proclamations of business acumen and a rhetoric of regional autonomy, they are outgoing, charismatic. They derive much of their allure from ridiculing their opponents in amusing displays enjoyed by their constituencies. Having long been looking to take back their forests from outside intrusion (politicians in the capital city, foreigners from the United States and Europe), the people of the region see the exporters as a potential path forward. They realize what the exporters offer is more charade than reality. They understand what they are embracing is the aura of independence rather than meaningful empowerment. But even fleeting symbolic victories are hard to come by for a people who have found themselves at the hands of oppressors for centuries.[6] Disgruntled residents, such as the rosewood workers in the northeast, serve as the so-called slum troops that remain a key feature in actualizing political turmoil in Madagascar.[7] Votes are bought in even the most advanced democracies, but in Madagascar, unrest is bought as well.

corruption is not the right word

It is tempting to label the political dynamics of rosewood in Madagascar corrupt. Bribes and legal confusion are not just pervasive; they form the backbone of the governance system. Rosewood policies—complex and contradictory—are intended to create an atmosphere of legal confusion in which few at the top can be deemed culpable for their actions. The central government has instituted an unpredictable dynamic of opening (*misokatra*) and closure (*mifody*), permitting thousands to enter the forest and driving them out just as quickly. Those tasked with stopping the trade bring chairs in which to sit along the river and impose fines on—rather than confiscate—the passing logs. Likewise, I am told that journalists come to the region in search of bribes *not* to write a story and are seen as "committing suicide" (*hamono tena*) if they partake in any actual reporting.[8] From the forest all the way to the capital city, rosewood has perpetuated this type of governance, which one might call corrupt if the word did not imply some sort of deviation from the norm.

But rather than an aberration, rosewood demonstrates the continuation of long-standing governance patterns that have persisted since Madagascar's first national leaders. Since the introduction of international currencies throughout the island, "all accumulators of money were potential political threats."[9] Even one of the most celebrated Malagasy heroes—the Merina king Andrianampoinimerina, who first began the political unification of the country in the late eighteenth century—was later revealed to be a merchant slave trader who had no political authority of his own but managed to use his earnings to purchase arms from the French and organize revolt. His first order of business upon securing rule was the monopolization of the most lucrative trade—slavery—in order to cement his own fortune and preclude that of his rivals. "The problem," anthropologist Maurice Bloch notes, "was that what had been possible for him was also possible for others."[10] Traders in Madagascar had become both the politicians and their adversaries: political authority was little more than a thinly veiled "attempt to pump off this loosely flowing power [money through trade] in order to reduce the possibility of challenge."[11] Rosewood—next to vanilla, the region's other lucrative export commodity, discussed in Chapter 2—was no exception.

As in many developing countries, the modern political system in Madagascar emerged alongside the country's integration into the global monetary system. From the late eighteenth century to the beginning of the colonial period in the late nineteenth century, those elite actors who controlled the most lucrative export markets held much of the country's political authority as well. After the annexation of Madagascar to the French in 1894, however, the colonial regime usurped control of nearly all resources on the island. The export of rosewood and other precious hardwoods was monopolized by the new colonial administration. Vanilla, too, was completely controlled by the colonial regime, which went so far as to strictly prohibit its cultivation outside the northeastern portion of the island. Beyond export commodities, the forest itself was—at least legally—in the hands of the colonial administration. As discussed further in Chapter 4, the French metropole via the Malagasy colonial state became the sole owner of forests in Madagascar, permitting only the most basic subsistence use rights to neighboring villagers.

Forestland in Madagascar was thus transformed into what political ecologists Nancy Peluso and Peter Vandergeest refer to as "political forests"—tools for nation building and consolidation of state control.[12] Far from being a universal or purely ecological category, the designation of land as official forest is a deeply political process that *makes* the forest, both materially and discursively, as much as it finds it growing out there in the world. Through the forest, the state transforms from a far-off appendage whose machinations remain aloof and inconsequential into a looming authority that lays claim to what might very well be understood as one's own backyard. In the northeastern forests, the colonial administration prioritized the export of rosewood and other precious hardwoods, exporting from the region tens of thousands of tons of hardwoods.[13]

But as geographer Philippe Le Billon and others remind us, the control of resources is not always a politically stabilizing pursuit. Forest resources in particular can fuel insurgency and rebellion just as easily as they might help build the official state. Rosewood demonstrated this clearly beginning in the early 2000s, when demand from China began to pick up and the Chinese Malagasy diaspora in the northeast was uniquely positioned to profit from it. These elites—not quite insurgents but in a precarious relationship with the state—leveraged their earnings to gain popular support and secure offices within Madagascar's newly elected government. Rather than building a post-coup regime, however, as the political forest thesis might suggest, this powerful new bloc destabilized the government from within. It resembled what political scientist William Reno refers to as a "shadow state"—a type of individualistic, commercially oriented governance "constructed behind the façade of laws and government institutions."[14]

As with Jean-François Bayart's *pays réel* (as opposed to *pays légal*), shadow states derive their power from the private control of resources exercised through "clandestine commerce" and illicit market transactions, often via elite global connections.[15] Elite sites of resource control within Africa are connected to wider global networks through a type of "enclave extraction."[16] At the expense of long-term state making, shadow states monopolize these networks to their exclusive benefit. This is the flip side of the political forest—not the mutual state and forest making of a nascent colonial regime but the hollow politics of a state that has abandoned all pretense of nation building.

The destabilizing effects of rosewood politics since 2000 have been perhaps most intense in Madagascar, but they are not unique to the country. Because of China's surge in demand, new rosewood shadow networks have cropped up across the globe. Elite traders in Myanmar, Laos, Cambodia, Thailand, and Vietnam all use illicit circuits to export endangered rosewood logs, gaining record profits.[17] In many of these countries, shadowy ties between villagers, timber traders, and state officials permit the clandestine trafficking of rosewood logs despite bans on the trade.[18] While initially focused in Southeast Asia, Chinese rosewood importers have increasingly turned to Africa to bolster dwindling supplies.

Gambia, Benin, Togo, Ghana, Mozambique, and the Democratic Republic of the Congo, among others, now export rosewood. Rosewood from Madagascar, however, both highly endangered and considered to be of superior quality, provides the most valuable African alternative and thus the greatest opportunity for shadow state politics.

While certainly not the sole provocateur of political tumult in Madagascar, rosewood made it to the top of the list during the past two decades. Burgeoning demand from China has bestowed upon an elite few profits that rival the state budget, allowing rosewood exporters to enter the government and monopolize the trade. Residents of northeastern Madagascar are well aware of this dynamic and not entirely opposed. This chapter will show how rosewood represents regional autonomy and self-determination for a constituency that feels exploited and overlooked by the central regime. As counterintuitive as it may seem, the rise of rosewood exporters within the central government imparts a feeling of regional ascendancy and, along with it, a sense of restoring the forest to the people. As symbols of potential prosperity, these exporters-turned-politicians have gained support from constituents all along the rosewood trail, from the inland villages to the coastal cities. After more than a century of being cut off from their forests in the name of empire or, more recently, conservation, people across northeastern Madagascar have turned to rosewood—and the rosewood exporters—as a way to take back the forest, even if only through hollow political victories that might end up doing more harm than good.

on the rosewood trail

Rosewood is logged from three main protected areas in northeastern Madagascar: Masoala National Park, Marojejy National Park, and Makira Nature Park (Figure 3.1). Logging is not permitted within these parks, but after the coup in 2009 thousands entered in search of rosewood regardless. When former president Ravalomanana finally resigned and fled the country, rosewood exporters threw an all-night party in celebration. The following day, trucks with speakers drove around town announcing new work in the forest. Radio ads and posters solicited young men with a sense of responsibility and adventure. Villagers living near the region's parks were enlisted to house the new workers and join in the work themselves.

Rosewood logging is arduous and dangerous work. "*Aza mivazaziaka,*" loggers tell themselves and each other. This is a blanket phrase used throughout the process of logging and dragging rosewood from the forest. It means many things:

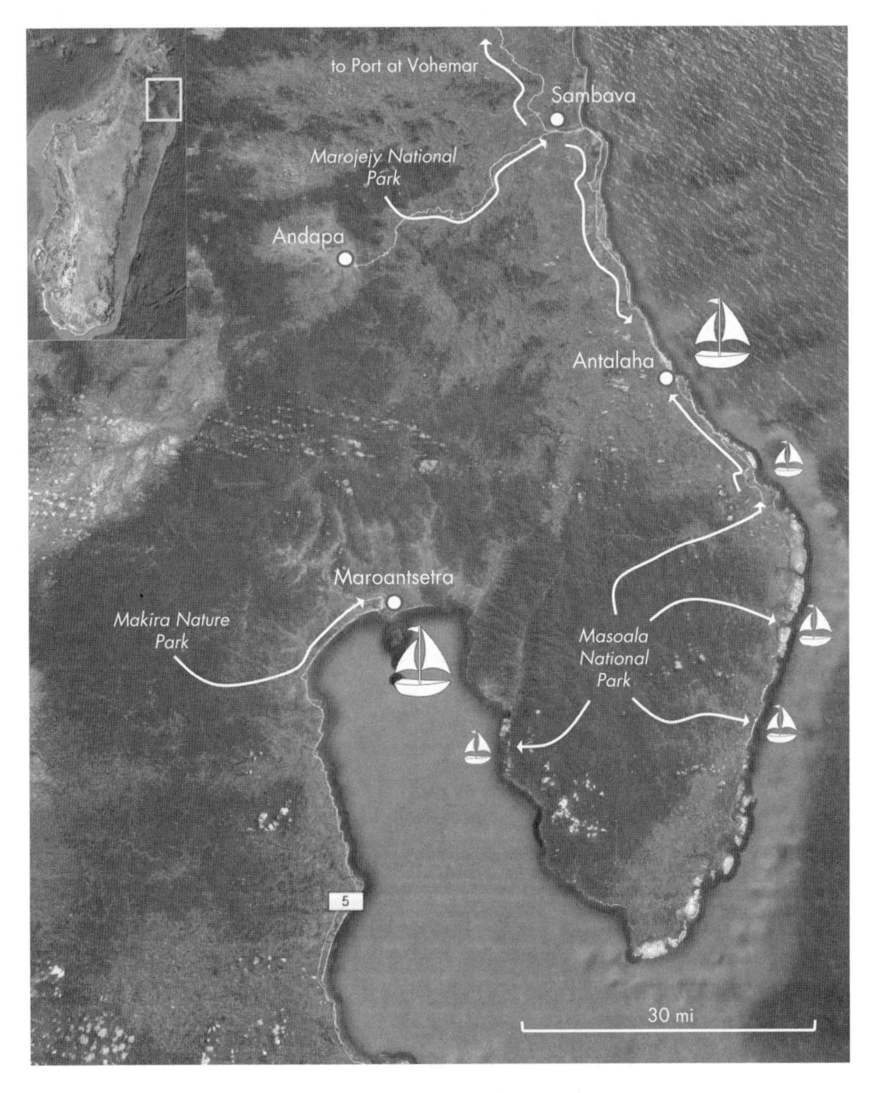

Figure 3.1. Overview of the rosewood logging and export routes. Trees are logged from their sources in Masoala, Marojejy, and Makira parks and then dragged, driven, or taken by canoe to boats arriving at the coastal cities and villages. Satellite image from Google Earth with overlay by Annah Lake Zhu.

don't screw up, don't be afraid, don't slip and fall, and above all don't be slow to jump if a log rolls furiously in your direction. The terrain is mountainous, slippery, and dense. Injuries and starvation are common; death is not unheard of. For the most part, as noted, the work is not mechanized. Rather than chain saws or heavy machinery, operations involve dozens of men harboring ropes and remarkably dull axes. Often deep in the forest, rosewood trees may be days from the nearest camp and sometimes weeks from the nearest weighing station, where the logs are bought by the kilogram and transported thereafter by canoe or car.

Rosewood logging operations officially begin when two scouters enter the forest in search of trees. In the early 2000s, scouters could find nearly ten rosewood trees per day. By 2015, one or two trees amounted to a laudable daily find. Sighted trees of respectable size and quality are marked through rudimentary preparations—a few jabs at the trunk, for example—meant to indicate to neighboring scouts that they are taken. Scouters return the next day to cut and prepare the trees. This involves clearing vegetation surrounding the trunk, felling the tree, removing the tangle of branches at the top, peeling the outer bark off the remaining trunk, and cutting the length of the tree into one- to two-meter logs. Following these steps—which might take anywhere from one to six hours for a single tree, depending on the size of the tree and quality of one's axe—a team of two people can prepare an average of six logs per day.[19]

Once prepared in this fashion, logs are deemed "ready to go" (*en carton*), and what is considered the *real* work begins. Scouters return to the prepared logs but now with a team of as many as sixty men to drag the logs by rope from the depths of the forest. The largest missions—involving larger trees sixty to one hundred centimeters in diameter, hidden deep in the forest—take one to two months to execute with a full team of sixty workers. Smaller missions—involving smaller trees thirty centimeters in diameter or less—last only a few days, with just two workers hauling the logs. An average log on the steepest slopes requires eight people, three on each side and two alternating at the front, each pulling a rope tied around a well-placed engraving made at the far ends of the log. The lead, referred to as the *lampy* (flashlight), is the most dangerous position. Steep slopes or rushing rapids could easily deliver the log in a fatal blow to the man at the front. Workers hurry to the ropes to avoid this position. Latecomers cringe when they realize that this is the only rope left unmanned.

Workers typically earn 10,000 to 15,000 ariary (US$3–$4) per day, considerably more than wage labor rates in vanilla and agriculture.[20] Logs are hauled through the forest to the nearest river, which is typically too small and intermittent to harbor a canoe. At the river, four rather than eight transporters are required, now dragging the logs through alternating deep and shallow waters. As the river widens, truck tires or a makeshift raft can be used to float the logs, and at this point only two transporters are required. Small huts are erected at key locations along the river to provide limited rations of rice to passing workers. Those arriving late find the pots empty and continue along hungrily. Occasionally, I have been told, workers are consumed by such extreme hunger that they search desperately in the forest for anything to ingest—a small raw cassava, sour fruit—anything to fill their stomach instead of the dirt they might find themselves loathsomely considering.

In addition to hunger, a host of other hazards plague the trail. River transport in particular enjoys a veritable lexicon of impending dangers. *Double voie*—where the river splits into two parallel white-water rapids before reuniting in a violent confluence—is the most feared. "Many people die there" (*tenga olo mamoly*), I was told in the same fearful words by more than one interlocutor. Similarly, "black lanterns"—deep yet undetectable whirlpools that ingurgitate passing logs, retain them for a fated "five minutes," and then violently expel them at an indeterminate location downstream—have occasionally commandeered the lives of workers. Also feared are crocodiles, mysterious forest sicknesses, and the infamous "goodbye father" (*veloma baba*), an especially steep ravine named for the remarks ceremoniously uttered before attempting passage.[21]

The *tribunal*—or "court of justice"—marks the last dangerous passing along one of the trails, where steep slopes end in a vertical drop. It is here, I am told, workers are inescapably given a sentence of life or death. "Life depends on luck" (*ny fiainana dia miankina amin'ny anjara*) goes a famous Malagasy saying. By indiscriminately ending the lives of some of its passers, the rosewood trail has become intimately connected with luck. The trail can be, at varying passages, a villain, trickster, or judge. Walking along it, loggers find themselves the protagonist to either outwit or be outwitted accordingly. Those who survive the final passage—who are ruled "innocent" by the forest's tribunal—pause to wipe the sweat from their brow and exhale a sigh of relief before continuing along to the camps.

in the camps and villages

Logging camps provide the nexus between the villages and the last remaining rosewood trees growing deep within the parks. Typically along navigable rivers, camps occupy relatively accessible areas to which supplies can be transported. After one or many days' work in the forest, loggers return to the camp hungry and tired. Depending on their working arrangements, food may or may not be provided by their bosses. If not, loggers patronize private vending stalls through a vibrant secondary economy that has sprouted up to support the rosewood trade. Just after a job is done, for example, they might rush to the nearest Malagasy bread (*mofogasy*) vendor—the one scoffed at days earlier for charging twice the going rate—and devour everything on the shelf without regard for price. Loggers might then sip some Malagasy wine (*betsa betsa*) before falling asleep on a pile of delicately folded branches under a makeshift cover of sticks and leaves.

After the coup in 2009, as told by my interlocutors, hundreds of camps were erected in and around the protected areas to accommodate the new migration of loggers searching for rosewood. Cooks established temporary stands serving rice and broth. Traveling merchants sold whatever they could haul with them into the forest. Women who arrived made quite a killing as hordes of newly paid men bid extravagant prices for their company. Music played day and night on ported-in speakers or the cell phones of crew bosses who managed the logging effort. Mattresses might also be ported in for the bosses, and loggers who were especially hardworking, clever, or amusing might be permitted to share in their comforts.

As the trade ebbed for reasons of weather or politics, the camps shed their layers of excess. When I traveled to a small village outside Masoala National Park as a Peace Corps volunteer in January 2011, for example, my entourage was greeted by the rhythms of Lady Gaga as we arrived by canoe. Climbing up the hill from the river to the village above, we were confronted by a row of recently emptied bamboo stalls lining the main path. Two giant speakers hooked up to a rusty CD player in the bamboo skeleton of an empty discotheque blared to a crowd that was not there. This all used to be filled with people, our guide told us. It was shut down after the military was sent in at the end of 2010, not long after *National Geographic* ran a feature on the rosewood logging crisis titled "Madagascar's Pierced Heart." Later we discovered that until a month before our arrival there had been nightly parties, but now there was nothing but the persevering speakers in a village largely deaf to its vibrations. That is, of course, not to say the logging stopped. As a consequence of the international outcry, the flagrant jubilation ceased, but the logs continued their journey down the river. We in fact saw a group of loggers ushering a shipment not long after our arrival—thirty-two logs sold for 3 million ariary (approximately US$1,500 at the time), we were told (Figure 3.2).

In the more remote camps, things can get dire. Along with supplies dwindling, stalls empty, the music stops, and vendors disappear. Rice is rationed and clothes are worn into rags. In a harrowing depiction provided by one of my friends who dabbled in rosewood logging, hungry men slept shirtless with tattered money stuffed in old plastic bottles tied at their necks. Plagued by aches, pains, and other mysterious ailments, they yearned for quick remedies. A single pill of medicine—*any* medicine—could be auctioned for quite a fee, with the unwell feverishly demanding remedies regardless of their intended purpose. My friend, who worked as a logger at the time, relayed to me a story of a man with severe backaches begging to buy stomach medicine from him. "I told him it would not help for his pain," my friend recalled, "that it was for the stomach, but he

Figure 3.2. Loggers transport rosewood logs through a village in northeastern Madagascar.
© Annah Lake Zhu.

was insistent." The pained logger continued to raise his offering price until my friend, not quite knowing how to respond, just gave him the medicine. With little else besides their wages hanging from their necks, loggers still deep in the forest throw what little money they have at any problems coming their way.

Finally, after enough logs are cut and delivered to the camps, they are dragged onward to the more connected villages, where they are weighed and purchased by mid-level traders at the behest of the rosewood exporters—to whom, of course, all the logs eventually flow. These mid-level traders (educated, urban, and scrawny compared with the muscular haulers) pay the crew bosses what is considered an enviable sum in the eyes of both the loggers they employ and idle onlookers too old or too weak to engage in the trade. The logs are then commissioned to be sent by truck or canoe to the larger coastal villages and cities that serve as key nodes along the rosewood trail. A number of villages on the Masoala Peninsula (including the village I traveled to in 2011 and afterward) are now considered "famous" because of rosewood, and many of their villagers are considered "rich" from the trade. The village presidents (*chef fokontany*), some earning, I was told, approximately 60,000 ariary (US$15, or close to a month's worth of sustenance) per batch of rosewood in transit, are certainly among the richest. With their wood houses, tin roofs, cold beer, and solar panels, these villages stand out from the others that have not managed to become hubs along the rosewood trail.

As with the camps, the villages are subject to the same oscillations of wealth and impoverishment. Small palm huts of those excluded from the trade butt against the crew bosses' larger tin-roofed abodes with logs piled high in the backyard. Recently paid workers might drink away their weekly wages in a single night's binge. Discotheques built and danced in during boom periods are abandoned soon after the money runs out. Impromptu casinos surrounded by crowds of loggers fresh from the forest disappear weeks later. Blankets spread out in the fields for migrant loggers are soon washed and folded and stored for future use. As has been documented in resource booms and busts throughout the country, the peak of the trade is mirrored by its trough.

in the cities

Historically wealthy from vanilla, the cities of northeastern Madagascar now receive a further influx of foreign exchange from the rosewood trade. Antalaha in particular has become known as the heart of the rosewood trade. The most notorious rosewood exporters hail from Antalaha, with a few others residing in the nearby cities of Sambava, Maroantsetra, and Vohemar (see Figure 3.1). Many of these exporters, as noted in Chapter 2, are of Chinese ethnic origins, having arrived in Madagascar during a wave of immigration that began in the early twentieth century. Cantonese speakers by birth, they have since come to learn Mandarin—"the language of money." As the rosewood trade began to escalate in 2000, all logs cut from the forest had to first pass through Antalaha before continuing their journey overseas to China. After the 2009 coup, however, Chinese ships filled with Malagasy currency began arriving all along the coastline to buy rosewood direct from the forest. In addition to the residents of Chinese ethnic origins, many new entrepreneurs began joining the trade.

Rosewood money continues to envelop northeastern Madagascar in a complex mix of opportunity and despair. As the rosewood trade escalated, residents witnessed dramatic changes in their surroundings. Large concrete houses sprouted up, motorcycles buzzed along the main roads, and beachside gatherings hosted by rosewood traders became commonplace. Many residents—although certainly not all—benefited from the new prosperity. School fees were paid on time. Young men came back from the forest with more money than ever before. When not squandered in a single spree, rosewood money bought cows or a computer, whatever goods that could be leveraged when the market inevitably came to a close. In just two months of work, some traders could raise enough money to purchase a motorcycle or build a house at the edge of the city. In a similar time period, loggers could afford a new bed or tin roof for their house.

But the trade was a double-edged sword. Residents not directly involved saw little change in their own incomes while the price of market goods soared. Taxi fare doubled. The city bus no longer serviced the main road, as it was reportedly repurposed into a truck for shipping rosewood. The region's ports became clogged with logs, creating a shortage of other domestically shipped commodities. Rosewood exporters laundered their money through other industries in the region—most notably vanilla—causing undersupply and market instability. Although parents could now pay their children's long-overdue tuition, many teachers deserted their students to profit as low-level traders. Just outside the city, agricultural labor also dwindled as young people in the countryside made their way to the forest in search of rosewood. What are the youth to do, I have been asked many times, when the trade stops and the money dries up?

The Janus face that accompanies any great influx of money has been the experience of the forests, camps, villages, and cities across northeastern Madagascar. While the rosewood exporters have collectively recorded more than US$1 billion in profits, others have spent their meager earnings in one egregious spree. It is this uneven dynamic that, as the following section will demonstrate, created the conditions for a new power bloc to emerge within Madagascar's Fourth Republic—not through a coup or dictatorship but through democracy.

the rosewood elite

The first spike in rosewood exports since the colonial period began after Cyclone Hudah in 2000, when the government issued a series of salvage logging permits for fallen trees. The cyclone destroyed the majority of rice and vanilla crops in the region, undermining both subsistence and income.[22] In its aftermath, a few individuals with connections to China petitioned the government to permit salvage logging for economic relief. With their petition approved, these individuals began to export large quantities of the wood to various international buyers—especially those in China. Shortly thereafter the central government reversed their permissions, issuing two ministerial orders prohibiting rosewood extraction and export in an attempt to temper the logging frenzy that ensued and appease international donors' growing concern.[23] According to residents, this reversal made little impact on the trade, which continued apace.

Another round of logging and export permissions were issued after another cyclone in 2004, which destroyed additional farmland and was accompanied by a crash in the price of vanilla. The twin disasters of the vanilla market crash and cyclonic damage left rural families with a gaping income gap exacerbated by food price hikes. Rosewood exporters once again petitioned the government and more salvage permits were issued, allowing loggers to return to the forest.[24]

Although boosting the trade, the salvage permits issued after the cyclones in 2000 and 2004 paled in comparison with the permissions given in the wake of Black Monday at the start of the coup in 2009. Two days after the Black Monday protests, soon-to-be-exiled President Ravalomanana signed a decree allowing thirteen individuals from northeastern Madagascar (many of the same as before) to export rosewood with impunity.[25] Soon after that, the newly installed opposition leader, Rajoelina, reaffirmed this legislation, facilitating a protracted logging frenzy of questionable legality that continued largely unrestrained for one year. This resulted in the export of more than one thousand shipping containers holding at least fifty-two thousand tons of precious wood, earning the rosewood exporters at least US$220 million in 2009 alone, likely much more.[26]

While logging and export continued apace in the northeast, making millions for the thirteen listed rosewood exporters, by the end of 2009 the recently installed transitional regime sought to secure its slice of the profits (see Figure 3.3 for a map of the financial flows of the trade). Crippled by international budget

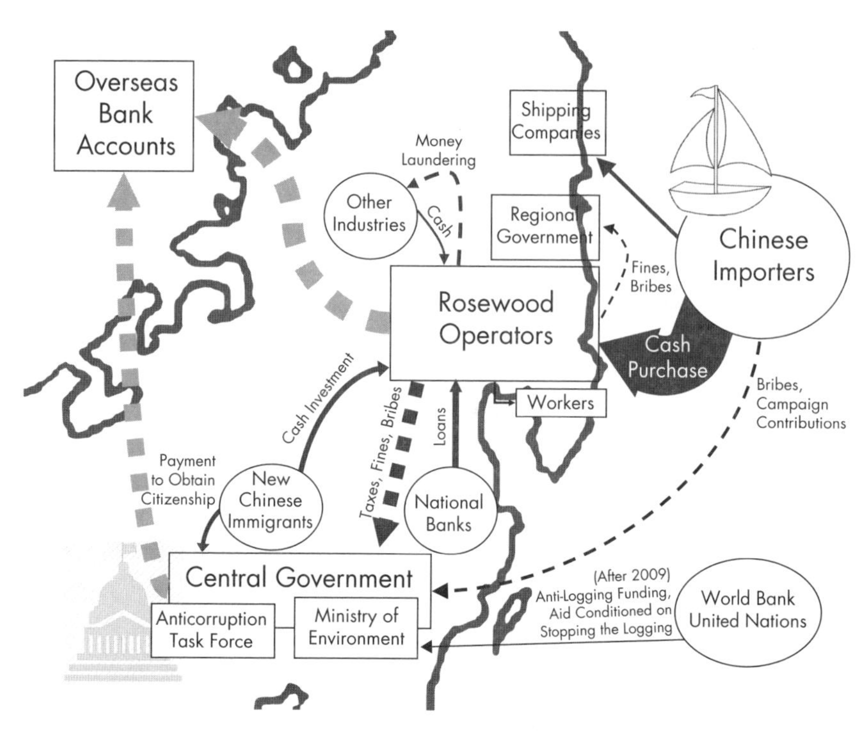

Figure 3.3. Financial flows and key players in the rosewood trade. Arrows approximate relative size and direction of financial flow. Dashed lines represent transactions of questionable legal status. Circles represent sources; squares represent sinks. © Annah Lake Zhu.

support cuts of nearly 90 percent following the cessation of aid after the coup, the transitional regime was in desperate need of finance. It permitted export to continue while benefiting from fines that amounted to 30 percent of the value of the trade.[27] This move generated short-term revenues of as much as US$40 million but outraged Western donors.[28] By 2010, international pressure to stop the trade forced the transitional regime to finally take action.[29]

"they only look for money"

After various superficial attempts, in August 2011 Rajoelina promulgated the most comprehensive anti-logging laws yet. In a landmark executive order, he prohibited all logging and transport operations, canceled all prior licenses and legislation, imposed heavy fines, and permitted little legal recourse for suspected offenders (see the left column of Table 3.1). For nearly six months following this

Table 3.1. Contradictory decrees to "clean up" the precious woods sector in Madagascar

President's Executive Order, August 2011 (No. 2011 / 01)	*Minister of Environment's Decree, January 2012 (No. 0741 / 2012)*
Prohibits the cutting, transportation, operation, marketing, and export of rosewood and ebony (Article 1)	Authorizes the export of rosewood and ebony in all forms (Article 4)
Cancels all export licenses in force (Article 3)	Grants export license to anyone who formally requests it, at the discretion of officials (Articles 5, 6, 7)
Cancels all previous contradictory legislation (Article 14)	Referring to earlier contradictory legislation in the preamble, insists that the decree will take effect "regardless of its publication in the Official Journal of the Republic of Madagascar" (Article 15)

Source: Reformatted from "La 'bolabolacratie,'" *Madagascar Tribune*, March 1, 2012.

decree, the conservation community released a tentative sigh of relief. But then, in complete contradiction, the minister of the environment issued a separate ministerial decree authorizing the ministry to distribute new export licenses at its discretion (see the right column of Table 3.1).

Through these contradictory decrees, the Malagasy government first appeased Western donors by implementing draconian restrictions and briefly halting the rosewood trade entirely, and then, six months later, consigned another branch of government to selectively reauthorize export permits at its own discretion. Thus, while appearing to stop the trade entirely, this comprehensive "cleanup" campaign instead ensured that only certain exporters—those allied with the central regime—could continue. Indeed, the Malagasy government "'cleaned up' the sector by taking full control."[30]

To enforce this and other attempted "cleanups," national troops were sent to the region to halt the logging. Residents are familiar with the chain of events accompanying the military's arrival and recounted them to me roughly as follows. First, troops are deployed, with the most elite bosses informed well in advance of their arrival. Radio broadcasts issued upon the military's arrival inform the others. Soldiers arrive first in the coastal city of Antalaha: six to ten big trucks of uniformed men with guns driving south toward the logging villages of the Masoala Peninsula. As they drive down the main road, onlookers glance over and know immediately that they are there because of rosewood. They watch the procession, mumbling to one another as the trucks pass, "They look only for money" (*ihany no mitady vola*).

Meanwhile, messengers are sent to the forest to further publicize the militia's impending arrival. Loggers quickly bury any logs in their possession and return to the villages to wait idly, pretending to be native villagers. Traders who have made recent purchases hire everyone they can find in the vicinity to hurriedly bury their stock. Then they too pretend to be idle villagers. The military, I am told, are not fooled by this ludicrous mimicry, but for a small fee they can be assuaged. A few traders, however—those who have not buried their stock in time or do not have the funds to appease the enforcers—have their logs confiscated, and they return to the city having lost their entire investment. An even smaller handful are actually arrested and delivered to the jail in Antalaha. When the military finally leave the vicinity a week or two after they have arrived, the remaining traders dig up their logs and resume operations.

In protest, the rosewood exporters have organized strikes against the military intervention. Over half a dozen strikes, I am told, have been held in the northeast, to varying effect. I have not witnessed these strikes, but they have been described to me by a number of interlocutors who participated. A typical strike begins when key rosewood exporters hold a meeting at the town hall, explaining the transgressions of the central government and inciting followers to join the strike. The strike leaders (the rosewood exporters or those hired by the rosewood exporters) speak of the government's abuse of power, its preposterous demand to stop the trade of the biggest source of revenue coming from the region. Making their point with ad hoc editorials, they facetiously speculate on whether government officials might stop the trade in potatoes in the capital city as well. They emphasize the absurdity of this narrow prohibition for this particular tree and no others. They are animated and entertaining. Strike organizers gather crowds from the city while trucks are sent to the countryside and village leaders are paid to send their residents to participate. An ox is killed in celebration. The town hall overflows with people either driven in or arriving by their own volition.

Building on the fervor of the meeting, the strike ensues. Protesters march with signs of discontent. "Go home military, we are safe here" (*mody miaramila fa aminay aty milomina*), the protesters write on large sheets of paper and chant in unison. Loggers, paid participants, and idle bystanders captivated by the message of regional independence all partake in the commotion. Rallies like this, I am told, occur three to four days per week for about two weeks. Rosewood exporters watch from afar, allegedly paying shop owners at the center of town to close their businesses as further testament to regional solidarity against the military intrusion.

The demands of the strike are simple: military, return to the capital, release those you have jailed, and let us sell rosewood! While many participate in the contestation, many more deem it a regrettable mess—at best inconvenient and at worst a deplorable confirmation of the power of the rosewood exporters over the masses. One strike organizer with whom I spoke laughed at the ease with which the youth participate, wondering to himself at the same time that he pocketed his cut from the day's activities: Why can't they see that all this rosewood just makes the rich richer and the poor poorer?

campaign promises

Leading up to the elections in 2013, military interventions were abandoned along with the strikes against them. Logging and export restrictions imposed by the transitional regime were again completely reversed, and a new logging free-for-all ensued. Rosewood exporters from all over the northeast courted voters with their newfound billions in an attempt to gain the coveted parliamentary seats from each of the region's districts. Candidates, I was told, rented planes touting their names. They threw weekly parties, set up rural stations distributing sheet metal and other building materials across the countryside, and even gave handfuls of cash to strangers on the street. They paid for students studying in the capital city to return to the region and participate in the propaganda. After their victories, they paid for the same students to return once more in celebration. The rosewood exporters' campaign promises were as fantastical as the tactics: free hospital access, a complete revamp of all public schools, and a road stretching from Antalaha to Maroantsetra—across prime rosewood territory. In their exaggerated displays, these candidates reminded residents that they were the gatekeepers of a market that had revitalized the region and now backed their promises of limitless prosperity.

For the most part, the tactics worked. Many of the elected parliamentarians from the northeastern districts were alleged to be involved in the rosewood trade, some quite deeply. On top of this, their favored presidential candidate, Hery Rajaonarimampianina, was elected president of Madagascar's Fourth Republic. One of the most reported-on additions to the new government was a new member of parliament, who was the first to be listed in the government order permitting rosewood export and is alleged to have been involved with the rosewood trade ever since its resurgence in the early 2000s.[31] He is also the same rosewood exporter who features so prominently in the region's stories of bone theft and Chinese-built moneymaking machines, as discussed in Chapter 2. Many residents of the northeast considered this newly elected parliamentarian to be the number one player in the rosewood trade—one of the first exporters initially involved and now one of the last exporters through whom all the rosewood must pass before leaving the country.

Interestingly, this particular rosewood exporter-turned-parliamentarian is somewhat of a rags-to-riches success story. According to neighborhood gossip, his father was lucky enough to study abroad in China, maintaining contacts afterward. The rising young exporter cleverly leveraged these contacts in the aftermath of the cyclone, notoriously becoming the wealthiest person in the region by the time of the 2013 election. In addition to financing his own campaign, he was reportedly one of President Hery's "principal sponsors," for whom he avidly campaigned and provided hefty financial contributions.[32] Moreover, along with a few others from the northeast, he formed a political party (the Union of Independent Deputies) with the newly appointed minister of the environment, who is alleged to have similar connections with the trade.[33] Together, these new members of government transformed from local strongmen to central political figures, forming a powerful coalition for clandestinely centralizing the rosewood trade.

The largest consolidation the rosewood market has yet seen occurred after the 2013 election. Just months after his installation in office, President Hery sent troops to the northeast, shutting down all but a small subset of operations.[34] When I arrived in the region a few months after that, thousands of loggers had returned from the forest. Rosewood exports, the returning loggers told me, were now restricted to only those circuits of the newly elected parliamentarians and their allies in the northeast. This extreme bottleneck caused the price of rosewood sold *within the region* to plummet, while the price of rosewood *at export* was rumored to have soared.[35] The result was monopoly profits for those within the central government still able to export their logs.

To this day, those involved in the rosewood trade remain in a legal gray area as a result of the series of alternating permissions and prohibitions that has characterized the trade for the last two decades.[36] Although this legal confusion has been significantly reduced since 2013, when the international trade in rosewood from Madagascar became prohibited under the Convention on International Trade in Endangered Species of Wild Fauna and Flora (CITES), various rosewood seizures abroad nonetheless demonstrated that shipments continued to flow overseas through clandestine circuits.[37] Moreover, in 2014, the minister of the environment (an alleged former rosewood exporter) traveled to the largest seizure sites in history (nearly thirty thousand logs seized in Singapore, amounting to more than US$50 million) and confirmed to port authorities that these logs were in fact shipped legally—an impossible confirmation, given the international trade prohibitions.[38] Because of conflicting messages from the Malagasy government, those responsible for the shipment were eventually acquitted in 2019 and all logs were ordered to be returned to them.[39]

Deliberate legal confusion and eventual acquittal is the undeniable trend with rosewood politics in Madagascar. Although Malagasy news outlets frequently cite accusations ranging from parliamentarians to other members of government to even the president, the accused all "seem to remain untouchable due to insufficient evidence."[40] As an alternative to incarceration, asset suspension is also difficult. After freezing the accounts of key rosewood exporters in light of suspicions of money laundering, authorities were soon forced to release them again following an outcry from the region. As noted in Chapter 2, most of those who export rosewood also export vanilla and employ a large proportion of residents in the northeast. Freezing the accounts of these export moguls, it turns out, has far-reaching consequences.

whispering unrest

Piled high in backyards, buried underground, and idly confiscated at the side of the road, rosewood stocks scattered across northeastern Madagascar serve as a constant reminder of as much as billions in forgone wealth. Loggers who have been long out of work hustle for odd jobs at the same time that they eagerly recall "the time of the rosewood." They perk up, as I witnessed, when rumors of potentially destabilizing political developments spread—the return of former president Ravalomanana from exile, the strike of the national airline. Now that the vanilla market boom is turning to bust (see Chapter 2), workers in the northeast are anticipating a rosewood revival.

The impeachment of President Hery Rajaonarimampianina in 2015 seemed to provide a potentially explosive opportunity. Exactly why Parliament decided to impeach its newly elected president remains unclear. While rising inflation and gross negligence were cited as triggering conditions, it seems that deeper political transformations incited this mutiny. The president had allegedly experienced a falling-out with his former alliances, some of whom included the rosewood exporters who recently entered office. Indeed, the most notorious rosewood exporter-turned-parliamentarian described earlier reportedly not only voted for the president's impeachment but also paid other members of Parliament to do the same, using what has become known in Madagascar as *vola miodina*—"rebel money."[41] Although Madagascar's High Constitutional Court overruled the impeachment as unconstitutional, Parliament deemed the court's ruling invalid, and loggers in the northeast waited with anticipation. After the June 26 Independence Day celebration, a former trader warned me that there would be another coup, reminiscent of the one in 2009. Although he did not want to reproduce the chaos of the coup, he was nonetheless preparing to make the most of it.

Despite Parliament's machinations and loggers' predictions, there was no coup the summer during which the president was impeached and subsequently acquitted. The political situation nonetheless remains shaky as stocks of rosewood logs around the region remain idle and overgrown with vines. Although these logs have become a part of the landscape—a makeshift playground or a place to dry laundry—at times their white noise seems to darken into a whisper. Indeed, through the mouths of disgruntled exporters shut out from the trade, Chinese importers warning that "export will continue no matter what, because their money can go through even the most highly placed doors," and thousands of loggers now out of a job, rosewood has been whispering unrest into the ears of anyone willing to listen for two decades.[42] With each new election cycle, the question of what to do with the region's billions of dollars' worth of cached rosewood logs—already cut and awaiting export—remains unanswered.

Madagascar's experience with the rosewood trade offers a window into how the electoral process plays out in a landscape of vast inequity. In a single election cycle, an elite few leverage their earnings in order to step into the national political arena and control the rosewood market to make even more. Here, the electoral process does not so much level the playing field but merely contours—even exaggerates—its highly uneven topography. The process is indeed characterized by corruption, but it is a form of corruption much less amenable to superficial reforms that target discrete practices such as bribery and favoritism. Instead, the democratic institutions that are themselves supposed to foster equality have been captured to sustain long-standing patterns of inequality.

Yet—despite the patronage, bribery, and exploitation—Madagascar's rosewood elite connect out-of-the-way places to the vast resources of the global economy. They are critical mainstays in the regional economy that cannot simply be cast aside. They have transformed distant Chinese demands for luxurious furniture into vast logging missions in the forests, bringing legions of Chinese ships to the undeveloped shores of northeastern Madagascar. For many living in the region, the rosewood elite provide a preferable alternative to exclusionary conservation laws, as will be discussed in Chapter 4, that have reigned since the colonial period. Their invitation to take back the forest is not so different from invitations made by leaders the world over to regain national control of resources in the face of a rising global economic tide that can be as suffocating as it can be liberatory. These invitations are easily cast aside as aberrant "corruption," but they reveal the type of post-truth politics that threatens to destabilize even the most developed nations. Just as Madagascar's hot money spending brings to the surface the most intense dynamics found at the core of the capitalist economy, the country's extreme electoral politics brings to the surface political dynamics found at the core of even the most advanced democracies. Democracy in Madagascar underscores the fragility of democracy across the globe: democratic principles are not inherent or uncovered but must be painfully constructed and reconstructed, even in countries that boast a more than two-hundred-year history.

4

worst-case conservation

On World Environment Day in 2015, I happened to be visiting a notorious rosewood trafficking village.[1] Every year on this day, villagers can expect a visit from the *lehibe ny ala*—the foremost forest authority in the district (translated literally as "the chief of the forest"). Two of my colleagues and I were staying at the village during the forest chief's visit. We noticed that in anticipation of his arrival, a few things changed. Lumberjacks along the river abandoned their work. Miners took the day off from gold mining in the park. Beer bottles were taken off store shelves, and a number of resident environmental offenders absconded to the neighboring forests for the night. After the forest chief and his men arrived, they held an impromptu meeting at the village center, discussing the repercussions of deforestation and soliciting two cups of rice and 2,000 ariary per head (about 50 cents, or what many might make in a day of agricultural work). The payment was a convenient mixture of a fine, threat, and tribute wagered at the community level. It tacitly acknowledged that the village was a key hub along the rosewood trade and solicited modest payment for the unspoken activities. If payment was withheld, alternative measures would likely be pursued.

The following day, after the officials left, everything went back to normal. Environmental offenders returned from hiding and beer bottles repopulated the shelves. Lumberjacks resumed their work at the river's edge and gold miners returned to the park.

For many residents of northeastern Madagascar, this is conservation: a fleeting transformation inspired by visiting authorities. Swidden agriculture, logging, and mining are for the most part illegal, which simply means that they garner a fine.

The lumberjacks who resumed their work at the river's edge after the forest chief's departure, for example, were relieved to hear that my colleagues and I were not also looking to impose fines as we passed. They said that park agents come every week or so to collect fines, and being caught next to this fallen tree would have cost them 200,000 ariary (about US$50). Residents who engage in swidden agriculture for rice production are also subject to fines imposed by the visiting forest chief and other authorities. To ensure fines are paid, I am told, authorities typically visit just before the rice harvest and threaten to burn the fields of those in hiding.

In villages that benefit from more lucrative commodities, such as rosewood and vanilla, fines are often imposed regardless of evidence of illicit activity. This explains the village-wide payment to the forest chief on World Environment Day. All villagers, regardless of their participation in illicit practices, were compelled to contribute. It also explains the village's hidden beer repository. The forest chief and his men, we were told by villagers, not only notoriously demanded complimentary bottles during their visits but also lurked at storefronts, seeking further compensation from passing customers who could afford to buy beer. Hiding the beer circumvented this rent-seeking practice. Whether these impromptu rents went back into official coffers or the forest chief's own pockets is unclear—although it seems likely to be a mixture of the two. In general, the knowledge that villagers have rice to spare and money in their pockets means that fines will be imposed whether conservation restrictions are obeyed or ignored.

From the conservationist's perspective, this is the worst-case scenario. The widespread belief that villagers will be penalized regardless of their personal adherence to the law means that environmentally detrimental practices will continue despite regulation. In fact, the mere existence of regulations attracts rent-seeking practices from any group with the authority to impose fines. Nowhere has this been more clearly demonstrated than in the rosewood trade. Federal police, for example, are eager to be stationed in the northeast, where rosewood logs are trafficked. In peak trading periods between 2009 and 2014, these federal agents allegedly set up stations all along the river, collecting 10,000 ariary (US$3) per log that passed. In the cities, they demanded steep payments from elite rosewood traffickers, and, in moments of extreme temper or inebriation, they have been rumored to kill those who refuse.[2]

Conservation organizations are well aware of the rent-seeking that environmental regulations often inspire. In an attempt to combat this dynamic, they have instituted an alternative approach. One can observe this in the recently created Makira Nature Park, not far away from the village described earlier. In Makira, the Wildlife Conservation Society (WCS; formerly the New York

Zoological Society) has designated a number of villages to collectively manage the land surrounding the park's core conservation area. WCS is considered the primary manager of the park, not the Malagasy government. The forest chief and other environmental authorities participate, but they are always accompanied by a WCS employee. Moreover, to further reduce the temptation to solicit bribes, non-WCS authorities are paid per diem for any work they must perform in the park. By design, nearly all interactions between villagers and forest officials are mediated through the presence of a WCS employee. Because of this oversight and financial compensation, I am told, there are far fewer bribes solicited here.

Makira is an iconic example of what many refer to as community-based or participatory conservation. Participatory conservation attempts to achieve conservation goals through the inclusion of those people living in the vicinity of the conservation area. In a direct response to what has been deemed "protectionism," "fortress," or "fines and fences" conservation, participatory conservation maintains that the people living closest to natural resources are in the best position to assist with their conservation.

This chapter examines the historical transition from protectionism to more participatory conservation methods in Madagascar, with a specific focus on how the rosewood trade has influenced recent dynamics in three parks: Masoala National Park, Makira Nature Park, and Marojejy National Park. These parks exhibit different approaches to the challenges of increased rosewood logging and trafficking, ranging from more participatory to more protectionist. My purpose in analyzing these various approaches to conservation is to demonstrate not only their differences but also their similarities. As the pendulum swings back and forth from protectionism to participation, to resurgent protectionism and back again, it is useful to keep an eye on what is *not* changing. A fortress mentality prevails, I argue, no matter how participatory the conservation efforts claim to be. Even though *including* local villagers, participatory conservation models—perhaps more important, and to the extent possible—*exclude* "corrupt" local and national power structures that might interfere with and undermine conservation efforts. They attempt to circumvent or insulate against this corruption rather than address it head-on. Although less publicized, participatory conservation's greater goal is less to increase participation in managing a particular area than to *transnationalize* that area, insulating it from the harsh realities of the developing country. Thus, even participatory approaches erect a fortress, but it is a fortress of a different kind. Participatory methods attempt to circumvent the rent-seeking practices of worst-case conservation through their own globally minded fortressing.

Creating an insulated space to both mold environmental subjects and circumvent an exploitative institutional framework is, I argue in this chapter, the more pressing goal of participatory conservation approaches that contrast with strict protectionism. Participation is not about empowering locals to manage their own land how they see fit (which rarely happens in practice); rather, it is about instilling in these locals an environmental subjectivity that aligns with that of the transnational community. It creates within local communities—who are normally so far from the transnational environmental discourse of the day—a new environmental awareness, a new subjectivity. As the pioneering political scientist Arun Agrawal notes, through this unique technology of governance, the environment and environmental goals begin to constitute "a critical domain of thought and action."[3] At least, this is the desired outcome. High-value export commodities such as rosewood complicate this subject making, rendering the classic fortress approach of direct control and coercion a more necessary alternative. But in the best-case scenario, community-based projects instill in their participants a new understanding of the value of environmental protection and new daily practices for valorizing the forest in these terms. This is the paramount objective of participatory conservation and it requires direct contact between transnational and local communities and the removal, to the extent possible, of the corrupt practices that might undermine such environmental subject making. In this reading, both protectionist and participatory methods create a fortress. The question is, what is being protected from whom?

In northeastern Madagascar, rosewood has greatly complicated attempts at such transnational fortressing (Figure 4.1). When the military is inevitably sent in to shut down the trade, the result is a return to the worst-case conservation scenario of imposing fines across the board. In times of crisis, such as the aftermath of the coup in 2009, the international presence disappears and, along with it, any pretense of participation. Conservation funding is instead used to finance task-force-style militia who are just trying to get their proverbial slice of the cake.

The sections that follow use the case of rosewood logging to examine how conservation in Madagascar has shifted over time and from one protected area to another, ranging from participatory approaches molding environmental subjects to military-based interventions attempting to control the trade. The next section examines the historical trajectory of conservation in Madagascar, from imperial authority through to colonial fortresses, socialist isolationism, and neoliberal participation. The chapter then explores participatory and protectionist approaches to conservation through the specific cases of Masoala, Makira, and Marojejy parks. The final section turns to the theory of participatory conservation, making explicit how the rent-seeking practices associated with rosewood logging have inspired a worst-case conservation scenario that participatory approaches attempt to circumvent through their own type of (transnational) fortress.

Figure 4.1. Rosewood growing in northeastern Madagascar, still too young for commercial harvest. © George Zhu.

precolonial to neocolonial conservation

The primary form of subsistence in Madagascar relies on a system of shifting cultivation based on rotational forest burning and clearing known as *tavy* (swidden, or "slash-and-burn" agriculture). Traditionally, converted forestland has been used for extensive production of rice, the main agricultural staple. Before outside authorities intervened, local cultural taboos dictated which forests were allowed to be brought under cultivation. Elaborate customary traditions for when, where, and what to burn governed local practice.

As long as local taboos are followed, Malagasy people generally view the transformation of forest into agricultural land in a positive light. Through forest conversion, they extend their family lineage, claiming land for future generations and achieving a type of self-realization through descent. In addition to being a widespread land management technique, swidden agriculture is a cultural mainstay in Malagasy communities living near the forest edge. It is often performed as a ritual, connecting contemporary farmers with the ancestors thought to inhabit the land.[4]

This positive association with the pioneering of new productive lands conflicts with centuries of wide-scale prohibitions on the practice. One of the earliest outside authorities to prohibit swidden agriculture was the imperial Malagasy monarchy of the Central Highlands—referred to as the Merina.[5] The Merina monarchy claimed ownership of all forests in Madagascar and instituted a set of forest prohibitions for local users. They prohibited tree cutting, forest burning, and fuelwood gathering, although it is unclear whether the prohibitions applied to all forests or simply those they maintained control over. In either case, their forest code was largely symbolic. It is likely that Merina authorities did little to intervene in forest practices far outside the capital—especially those in the northeast. After the country's colonization by France in 1896, comprehensive forest regulations were established according to the French forestry model.[6] From the perspective of those living in northeastern Madagascar, the oppressors shifted from the Merina to the French.[7]

colonial fortresses

Madagascar's colonial forest service was created in 1900, and the country's forest codes, claiming all forests and trees for the state, were passed in 1913.[8] Vast tracts of land were designated as concessionary land for export-driven mining, plantation agriculture, and logging.[9] In general, the colonial regime greatly prioritized extraction and production for export over broad and balanced local economic development.[10] The export of precious hardwoods, including rosewood, palisander, and ebony, was the primary goal of colonial forestry in Madagascar. Precolonial timber export operations were greatly expanded and intensified, resulting in the export of tens of thousands of tons of hardwoods from the northeastern forests alone.[11]

In light of such extensive timber export operations, the colonial regime grew increasingly concerned about the conservation of resources and the adverse effects of erosion. Forest plantations of eucalyptus, acacia, and pine were established by the colonial state, covering one million hectares by 1928.[12] Alongside these plantations, the colonial state established land for pure conservation and scientific goals. By 1954, a total of ten million hectares of primary forest had been appropriated by the state, 9 percent of which was designated specifically for conservation and open only to French scientists.[13] The following year, the state introduced the idea of classified forests (*forêts classés*) to ensure even greater timber supplies. By 1959, classified forests added more than three million hectares of forestland to the ten million already appropriated by the state.[14]

Most important, the state banned without exception across the country the primary form of forest management prior to colonization: swidden agriculture through *tavy*. Malagasy people garnered only the most basic rights of subsistence use of forest resources. Taxes on all assets—rice fields, livestock, market transactions—were specifically designed to force farmers away from subsistence activities and toward export production and wage labor. This transition was portrayed by the colonial regime as the *mise en valeur* (valorization) of Malagasy nature and labor.[15] From the vantage of those living in the northeast, it was a more forceful continuation of Merina imperial policies.

In terms of the protectionist / participation pendulum, the colonial state promulgated a strictly protectionist approach. Conservation and development were in no way confused. "The French colonial government," political ecologist Catherine Corson notes, "addressed the ongoing conflict between production and conservation objectives by spatially separating the two," allowing either exploitation or conservation to exclusively define vast tracts of forest.[16] Nature reserves were proposed to be "equivalent to the forest [timber] reserves, but with a completely different purpose."[17] From the colonial vantage point, conservation and scientific goals could be met only if local communities were stripped of all usage rights.

In general, forest management and conservation in colonial Madagascar were considered purely a state concern—a science fit only for those with an academic degree. Contact with the forest outside of waged labor was prohibited. Malagasy people were characterized as "behaviorally unfit for skilled forestry work but adequate as brute man power."[18] In the eyes of the state, Malagasy management practices appeared "unplanned, aimless, nomadic, unproductive, and uneconomical in the utilization of land and labor and destructive of the environment."[19] Colonial management, in contrast, which prioritized the use of the forest for export, mass infrastructure, or conservation, represented unqualified "progress" as defined by the colonists.

Although the goal of colonial forest management was allegedly "progress," the result was the alienation of Malagasy people and their deeper movement into the forest. Rights-based access was restricted, requiring the state to be the formal intermediary of all major forest acquisitions. In practice, however, previous access patterns continued as well, albeit illicitly in areas of weak state control. To maintain access, many Malagasy communities traveled deeper into the forest, some even burning forested land as a means of state protest. Swidden agriculture as a form of subsistence also continued, although now illegally in areas removed from state authorities.[20]

By the end of the colonial period, the state had significantly alienated most Malagasy people living in the northeast. Malagasy nationalists reclaimed control of the country in 1960, thus marking the country's independence and the beginning of Madagascar's First Republic. Throughout the First Republic, however, ties with France and colonial legislation remained strong. Land tenure laws remained the same, and the state continued to claim ownership of the forest. Madagascar was still received in the international order largely through its colonial link with France and not as an independent economy. Forest exploitation and conservation patterns also remained largely the same, although export operations declined—especially timber exports from the northeast. This decline in exports became even more pronounced after socialist policies took hold of the country in the decade following independence.[21]

socialist isolationism

Although more in rhetoric than reality, the nationalist revolution in 1972 brought an end to Madagascar's First Republic and the colonial status quo. A consequence of growing peasant agitation, the revolution began with farmer protests in 1971 and culminated in the resignation of President Philibert Tsiranana in 1972. Following a few years of political turmoil, Madagascar's Second Republic—a unitary, one-party socialist state—began in 1975. The economic recession in the early 1970s paved the way for a new approach to development in Madagascar. The socialist state continued to claim control over land and resources but with different development objectives in mind. Socialist platforms steered away from the colonial mainstays of exports and conservation.[22] Resource controls in rural areas were largely unenforced, and villagers were encouraged to clear and develop land through subsistence-based agriculture.

The socialist period demonstrated a sharp rejection and rhetorical movement away from the Western discourses of conservation and forest economics altogether. In practice, however, the strategic absence of an alternative approach to forestry allowed existing forest codes to remain unchanged despite their rhetorical attenuation. Forest policy in general was avoided because of the colonial tensions it revived. Instead, agriculture was upheld as the main development objective, a domain that the colonial state had allowed trade to greatly overshadow. The separation of economic and conservation forest zones, however, continued as a colonial era holdover.[23]

Although issues of forest management specifically were often avoided, a general political trend of decentralization was embraced by the socialist state. The government's goal was to make the village assembly (*fokonolana*) into "the most basic organ of a socialist and democratic state with a view to putting development into the hands of the people."[24] According to the Malagasy president in 1977, the *fokonolana* functioned as direct democracies in which "the citizen's views are solicited every single day . . . a decentralized society . . . [which] gives broad responsibilities to all."[25] Hypothetically, general trends of political decentralization and the growth in power of the *fokonolana* allowed for a more participatory approach to forest management. There is little evidence, however, of forest access patterns actually changing during this time, except for a clear decline in export-based wage labor.

Ultimately, socialist reforms were expensive and reaped few financial returns. With the peasants still unappeased and agitation in the urban proletariat growing, the socialist isolationist state of Madagascar ended in the 1980s when, as with many other developing countries at the time, near economic collapse forced a severely weakened Malagasy government to accept a structural adjustment plan sponsored by the International Monetary Fund and the World Bank.

neoliberal participation

With the acceptance of a new wave of neoliberal aid packages, post-socialist Madagascar was reintegrated into the world economy through the global trend of neoliberalism rather than its colonial link with France. Conservation and development were placed front and center in Madagascar's new neoliberal trajectory. The turn toward participation was demonstrated very clearly in this neoliberal transition. Participatory approaches to forest conservation were legislated through various mechanisms throughout the 1990s, primarily with the aid of the World Bank and global conservation nongovernmental organizations (NGOs). These donors invested at least US$450 million in Madagascar's environmental sector between 1990 and 2013, resulting in the expansion of terrestrial protected areas from 3.2 million to more than 10 million hectares.[26] This, in turn, generated a plethora of new local management structures bolstered by transnational support. Practically every new participatory conservation technique developed at the international level has been reflected in Madagascar's domestic policies. Integrated conservation and development projects, community-based forest management, community-managed sites (KoloAla), and grassroots community organizations known as *communauté de base* (COBA) in French or *vondron'olona ifotony* (VOI) in Malagasy have all been developed in an attempt to achieve more participatory conservation.

This neoliberal turn toward participation in the 1990s was vastly different from the socialist decentralization two decades before. Both involved a rhetorical delegation of management to local users, but neoliberal participation often involved decentralizing both *down* to the local level and *out* to the transnational level. The state, though officially present, became somewhat of a vacated body, existing primarily as a vehicle through which the international community could claim land and organize those within it. Until the mid-1980s, the territorialization of Madagascar's forest was "primarily a project of rulers," but with revived support from the international community Madagascar has since experienced a specifically neoliberal territorialization under which access is determined across many scales.[27] The state participates, but it does so from the sidelines. Political ecologists Jim Igoe and Dan Brockington refer to this type of conservation as the creation of "transnationalized spaces, governed according to the needs and agendas of transnational networks of actors and institutions."[28] As international NGOs attempt to work with local Malagasy communities directly, the state has become increasingly marginalized. This has not been accidental but strategic. Local participation offers a clear route to circumventing broader patterns of "corruption."

Madagascar—experiencing periodic coups d'état and a notoriously nepotistic governance structure—has been particularly subject to such transnationalization. The key players are not foreign governments per se but transnational entities. The World Bank; the International Monetary Fund; the United Nations Environment Programme; the United Nations Educational, Scientific and Cultural Organization; the United States Agency for International Development (USAID); and the French Agricultural Research Centre for International

Development (CIRAD)—along with environmental NGOs such as Conservation International, the Wildlife Conservation Society (WCS), the World Wildlife Fund (WWF), the Missouri Botanical Garden, and Fauna & Flora International, sometimes working with large mining corporations such as QIT Madagascar Minerals (QMM, majority owned by Rio Tinto) and Ambatovy (majority owned by Sumitomo Corporation)—all have a say in how Madagascar manages its forests via strategically delimited conservation territories in various parts of the country. These transnational interventions aim to circumvent what many on this long list of actors refer to (publicly) as state capacity constraints and (privately) as rampant corruption. The territorialization of Madagascar's most recent protected areas, for example, was established through several working groups that involved a mixture of foreign aid donors, consultants, scientists, Malagasy government representatives, and NGOs. It was agreed that the management of these protected areas would be pursued by NGOs, the private sector, and local communities, all in addition to the state. Plans for designating these protected areas began with the Durban Vision under Madagascar's Third Republic, but it was only after the latest coup in 2009 that the transitional government signed these plans into law, nearly doubling the country's conservation lands through the codification of 171 new strictly protected and sustainable use sites covering a total of 9.4 million hectares.[29] The timing of this drastic addition was surprising, given that the country had not yet returned to electoral politics.

With the introduction of transnational networks and transnationalized spaces—in short, with the transnationalization of the Malagasy forest—forest access has been more mediated than limited. In many places, local access has been increased and encouraged, albeit strategically. During colonialism, independent forest access was largely outlawed, requiring locals to enter the forest either licitly as wage laborers or illicitly outside of state control. The turn toward participation has introduced more stakeholders to forest governance in Madagascar and complicated resource access. Instead of confronting black-and-white sanctions at the forest edge, foresters now confront a medley of stipulations and terms of use that they themselves may even have had a hand in devising. Structures constraining and molding appropriate forest use are not so much rights and prohibitions—everyone has a "right" to the forest, from its direct inhabitants to the entire global community. Increasingly, the terms of access are multitiered management plans and sustainable use blueprints made to appease the full range of stakeholders, from local forest users to the entire global community. These terms of access certainly involve local people in their establishment, but many elements of effective control over the land lie securely within the transnational arena.

rosewood complications

The proliferation of rosewood logging since 2000, and especially after the coup in 2009, has complicated the neoliberal turn to participatory conservation in northeastern Madagascar. Three parks in particular have suffered most from the logging: Masoala National Park, Marojejy National Park, and Makira Nature Park (see Figure 3.1). These parks are the site of the majority of illicit rosewood logging today.[30] Even though the creation of these parks was intended to bolster local management of forest resources, in many ways their exclusionary policies have continued the colonial tradition of expropriating resources from local control.[31] Rosewood logging has brought these dynamics to the fore. The experience of each of these parks during the rosewood logging boom after 2009 made it clear that their participatory approaches to conservation rely not on fortressing the forest from the people but on fortressing a transnational space of intervention from the rent-seeking practices of state and local officials charged with enforcing conservation laws.

Masoala National Park

Covering approximately two-thirds of the Masoala Peninsula in northeastern Madagascar, Masoala National Park contains the largest remaining contiguous block of tropical humid forest in Madagascar. It was established as Madagascar's eighth and largest national park in 1997 and is comanaged by WCS and Madagascar National Parks. Since 2019, Masoala (and Makira, as discussed in a later section) also receives funding from a US$22 million USAID contract awarded to the global consulting firm Tetra Tech.[32] Masoala covers a total of 2,300 square kilometers—about half the size of Yellowstone National Park in the United States. The park is composed of a core protected area that ranges from coastal forests at the edges of the peninsula to mountainous terrain at the interior, along with several smaller detached marine and terrestrial parcels. Surrounding the park, more than eighty thousand Malagasy residents share the Masoala Peninsula.[33] They engage in subsistence rice farming through permanent and shifting cultivation and grow vanilla, coffee, and cloves for export. After the coup in 2009, many residents also either logged rosewood or provided services for loggers in and around the park.

Masoala was established as an integrated conservation and development project (ICDP), meaning it was intended to achieve both conservation and development goals by creating a core protected area and also encouraging community-based management in a multiple-use buffer zone. The park was designed "in consultation with people at the local and national levels" and in an attempt to avoid conflict with village populations.[34] While priority was given to habitat protection for endangered species, local villages were mapped and excluded from the protected area borders where possible.[35]

The "development" part of the Masoala ICDP turned out to be minimal, amounting to a couple hundred dollars per year financed through a portion of visitor entrance fees to the park.[36] After a few years, the ICDP was abandoned entirely and Masoala became just a protected area. While the initial design of the park assessed the economic benefits of establishing sustainable forestry for timber export, calculating an annual gain of US$130 per household as compared with the business-as-usual scenario, actual implementation of such a sustainable forestry export market was never attempted. Instead, the hypothetical potential for sustainable forestry was used to justify the size of the core protected area, demonstrating that there was sufficient land outside the protected area for livelihoods.[37] Unfortunately, no actual plans for establishing community forestry for export or any other "community-based economic incentives" were discussed in the report. Likewise, no assessment was made of the practical feasibility of establishing a sustainable forestry export market.

Within this context, it comes as little surprise that the creation of the park has been met largely with apathy or aggression by local communities. Stating the matter very clearly, ethnographer Eva Keller notes that communities near the southeastern borders "almost universally perceived Masoala National Park as a threat to their livelihood, and only in the rarest of circumstances did anybody voice any opinion in favor of the park."[38] In contrast, another study in the northeastern region of the park found that villagers' sentiments ranged from general confusion over the park's existence to general support.[39] Many of those in support, however, lived farther from the park and believed that its boundaries were more prescriptive than permanent and that they could be adjusted for future use.[40] It was assumed that as the population grew, the park would be made smaller to accommodate increasing agriculture. Residents also believed that one of the park's main objectives was still development assistance, which actually had ended in the early 2000s after the park discontinued its development component. In my discussions with villagers living outside the park, the main complaint was not the park itself but the power the park gave to passing officials to solicit fines. Every few weeks, villagers could expect a visit from one conservation authority or another who would impose fines in cash or in kind for swidden agriculture or logging (Figure 4.2). The understanding was not so much that the authorities were being corrupt as that this was simply how conservation works.

The rosewood trade, of course, intensifies the dynamic. Masoala has been hardest hit by rosewood logging.[41] When the trade is "open," villagers living along the borders of Masoala work in the trade—either directly, logging and trading, or indirectly, providing housing, goods, and services. A survey of villages in the region showed that 27 percent of village households were directly involved in the trade and an additional 31 percent worked indirectly for the trade.[42] In certain villages that manage to become key hubs along the rosewood trail, the numbers are much higher. In the rosewood village that I visited, I was told that at peak times during the trade, all the houses were filled with loggers, and blankets were spread in the fields for additional loggers to sleep outside. Parties, as noted in Chapter 3, were held nightly. Logs were cached everywhere: backyards, riverbanks, canoes. Guardians were paid 5,000 ariary (about US$1) nightly to protect them. During slow periods, logs were buried in giant trenches, again with paid guardians standing by. Respondents estimated that at the peak of the trade, two hundred to three hundred boats per day traveled from the park along the Onive River to the coast. Chinese ships arrived all along the Masoala Peninsula to pick up logs fresh from the forest.

Along with loggers and traders, the rosewood boom attracted the authorities. Gendarmes and military stationed in the region brought chairs to the riverbank, collecting 10,000 ariary per log that passed. Local officials, too, sought their slice of the profits. The forest chief, I am told, was particularly strict, extracting fines from the entire village for any sort of environmental transgression performed—from rosewood logging and transport to swidden agriculture. If not cash, they demanded rice, often selling it back to the same group of villagers from which they obtained it. When the trade closes or slows, people in the region seek other economic activities. Gold mining in the park resumes. Swidden agriculture continues, probably at a more intensive rate to make up for rice that cannot be bought with wages.[43] Less lucrative tree varieties are logged and sold locally. Remaining rosewood scraps forgotten or fallen in the river are also collected for use and sale. On our visit to the logging village, we encountered a man lugging a small piece of old rosewood he had found in a stream. He, too, expressed his relief that we were not looking for payment.

As the rosewood trade began to intensify after the coup in 2009, the government organized an emergency task force to "secure" Masoala and the other protected areas in the region. Yet those members of the task force sent to the region were largely complicit with the dynamics of the trade. They saw their mission as

Figure 4.2. Lumberjacks near Masoala National Park. © Annah Lake Zhu.

not to stop the trade but to impose fines. For example, at a town near the park, rather than stopping shipments of rosewood logs at a checkpoint station, task force members charged 20,000 ariary per passing log as a nonnegotiable "toll fee." Some members were reportedly "very proud" of their role within the task force, acknowledging that "everyone needs to get their slice of the cake in this business."[44]

This sums up the story of conservation and logging in Masoala National Park rather well: everyone getting their slice of the cake, with those living closest to the forest often receiving the smallest slice. Although established as an ICDP, Masoala ultimately delivered minimal development benefits, which were provided primarily to more wealthy residents. Villages were not particularly active in establishing park boundaries, and community "consultations" typically meant conveying park boundaries after they had already been developed. In some cases, however, the boundary changed to encroach further on productive lands. Thus, productive activities now continue in certain areas of Masoala in a muddled, semilegal manner. The rosewood trade further exacerbates the dynamic, transforming the national park into a hotbed of rent-seeking behavior.

This is the worst-case conservation scenario. In fact, it is not really conservation at all but merely the situation that prevails when conservation restrictions exist alongside little else. Participatory conservation is intended to prevent this dynamic, to bring those living closest to natural resources more deeply into their management under the oversight not of corrupt local officials or armed militia but of the global intergovernmental and nongovernmental community. Corruption in Madagascar and other such countries, as noted in Chapter 3, is not an aberration: it is intrinsic, simply the way of doing business. Fines are not intended to prevent activities but instead serve as an informal tax system. This situation likely prevails in most protected areas that harbor lucrative export commodities such as rosewood. But it has been countered by participatory conservation methods, to a certain extent and with certain trade-offs, in Marojejy National Park and Makira Nature Park—the other two protected areas where rosewood still grows. The goal of these more participatory interventions is not to tackle corrupt dynamics directly but to create a space that is insulated from their disruptive effects.

Marojejy National Park

Marojejy National Park covers 55,500 hectares of mountainous land, about eighty kilometers north of Masoala National Park. Marojejy is much smaller than Masoala, but it contains greater altitudinal variation, transitioning quickly from lower-elevation rain forest to mountain peaks reaching an elevation of 2,132 meters. The park began as a strict nature reserve in 1952 after an eminent French botanist devoted a book to describing the region's natural beauty. In 1998 it became a national park, managed by Madagascar National Parks with partial assistance from WWF and, later, the Duke Lemur Center.

Marojejy is very different from Masoala in geography and management. As a much smaller park, Marojejy has a single entrance just off the main road from the coast to the city of Andapa (see Figure 3.1). The park has fifty-two communities living at its edges.[45] Before it became a park, two households lived within the interior of the reserve, but they were evicted when WWF became involved in 1993. In general, from the 1980s until WWF involvement, local villages engaged in land-clearing activities within the borders of the park in order to develop coffee and vanilla crops, facing few repercussions. WWF began campaigns to mark the edges of the park and establish forest surveillance programs, which helped maintain the park's integrity. When Marojejy became a national park, the boundaries were renegotiated, giving villages in the western portions more land but removing land from villages in the northwest.

Park relations with villagers in Marojejy are typically portrayed in a more positive light than is the case in Masoala. This is likely in part because there are fewer ethnographic studies of people living around Marojejy—my own conversations were also limited mainly to park and NGO employees rather than villagers—and also because of the park's smaller size and more centralized management. Marojejy has a more robust system of associations, including a guides' association, a porters' association, a cooks' association, and a women's association. "*Mafy ny fiaramiasa*" (strong partnerships), one employee noted when describing the park's relationship with surrounding villagers. The park has purchased school kits for local students, and every two to three weeks the best students from the villages are invited to visit Marojejy. The Duke Lemur Center also plays an active role, having established a community-based initiative in the region in 2012. Activities include environmental education, park border demarcation and monitoring, reforestation of fast-growing endemics and fruit trees, fish farming, and restocking of local rivers with endangered species.[46]

Marojejy, however, is not without its controversies. Limiting swidden agriculture both inside and outside the park is a constant struggle. Park boundaries have been (illicitly) moved physically inward by villagers, sometimes with the help of park employees. In one case, a park agent reportedly sold nine hectares of parkland to a local farmer, who then cleared the land for rice cultivation. In addition to swidden agriculture, bushmeat hunting, honey extraction, and selective logging illegally occur within the park.

Rosewood logging has also proceeded quite differently in Marojejy from that in Masoala. Because of its proximity to a paved road, Marojejy was hit first by the wave of loggers that surfaced after the coup in 2009. Loggers combed the park in search of rosewood and threatened to burn down the house of the park director if he impeded the operations.[47] Marojejy was shut down for three months amid the violence. When certain villagers mobilized against the logging interests, armed traders allegedly fired automatic weapons above their heads to disperse the crowd or threatened specific protesters with beheading.[48] As noted, the road to Marojejy was said to have been streaked red from a constant procession of logs dragged from the park to trucks waiting nearby. During these three months, six to eight trucks per day transported two to three tons of rosewood each, amounting to about fifty to one hundred logs per day.[49] Trees were cut primarily in the northeastern region of the park, where rosewood grows more frequently because of the humid climate. By May 2009, the park director, with the help of the military and local authorities, finally stifled the majority of the logging, and Marojejy reopened.

After the reopening, rosewood logging continued but at a much reduced level. Unlike the rosewood in Masoala, rosewood in Marojejy does not grow very far from the main road.[50] This, combined with village monitoring for logging in the sectors where rosewood grows, has made the logging much easier to control. As one park employee told me, park personnel have arrested loggers more than ten times since 2009, typically arresting groups of more than eight loggers at a time. On almost all of these occasions, a report from a villager triggered the arrest. After receiving a report, park employees (guides, porters, whoever is available) seek the help of additional villagers to gather a search group of nearly one hundred people. They wait until nightfall, I am told, and enter the park in smaller groups in search of the loggers. Just before leaving to apprehend the loggers, park employees may notify the gendarmes. They wait to provide notification until as late as possible so that the gendarmes cannot warn the loggers in advance, as has been said to happen, given their alleged connections.

The loggers are apprehended and removed from the park. If the gendarmes are present, they take the loggers to jail in a nearby city and confiscate the rosewood logs. "They do their formality and leave," a park employee I spoke with observed of the local and federal authorities. "The police and military are just a formality and always for the money," he continued. "You know how it is here in Madagascar." If found guilty, the loggers are incarcerated for six months to five years, I was told by another conservation agent. "If they pay," he then added, "maybe one to six months." Others pay immediately and do not go to jail at all. Protesting this type of bribery, the park director of Marojejy was rumored to have spent two weeks sitting next to a jail in which loggers had been recently taken in order to ensure that they would not be able to pay their way out. The management at the jail told him to leave, but he refused.

If gendarmes are not present, park employees act on their own. After apprehending the loggers, they cut the confiscated rosewood logs into small pieces to burn as firewood, not trusting the gendarmes' confiscation. They punish the loggers by taking off their shirts, covering them in mud from the rice fields, and parading them down the main road. Following this display, the gendarmes are contacted to take over.

Village surveillance seems to work well in Marojejy. I was told that the park, unlike Masoala, still has a large number of rosewood trees as large as sixty centimeters in diameter. "People are afraid to cut rosewood," the park employee noted. "We are in collaboration with even the children . . . everyone comes here to tell us when there are loggers." The gendarmes and police have a smaller hold on things here; they do not ask for as many bribes in the villages. Their connections, I am told, are limited to only a few families who have young men who work as rosewood loggers. All of the big actors in the rosewood trade—the exporters and the higher-level traders—are not from around Marojejy but rather are from the coast. Another conservation agent at WWF attributed this to ethnic differences. A bit more inland, residents around Marojejy are mostly Tsimihety, while those on the coast are Betsimisaraka. He recited the stereotype that the Tsimehety respect rules and traditional customs better than the Betsimisaraka. WWF helps them, he observed, and they get many advantages. The Betsimisaraka, in contrast, are coastal, more educated, more connected, less in need.

When I told Betsimisaraka coastal residents what I heard at Marojejy, they had a slightly different take on the matter. Yes, they agreed, there is more collaboration with villagers around the park in Marojejy. Unlike in Masoala, villagers surveil the park and report transgressions. But they do this, the coastal residents insisted, because they themselves have fewer opportunities to log. The real reason that Marojejy is not the same logging hub as Masoala, they maintained, is not less corruption or more collaboration but simply geography. In order to get the logs from the park to the boats at the coast, they must use the road. And it is a long road. Bosses would have to pay many, many authorities stationed along the road to proceed with their logs to the coast. Why bother, when Chinese boats can come directly all along the shores of the peninsula and buy rosewood straight from the forests of Masoala National Park?

Makira Nature Park

Constituting a core protected area of 372,470 hectares, Makira replaced Masoala as the largest conserved forest in Madagascar after its formal establishment in 2012 (launched in 2001). The two parks are very close together and nearly touch in some areas. Together, they provide the largest protected area of contiguous lowland tropical forests on the island. As with Masoala, Makira is a joint venture between WCS and the Malagasy government, but in the case of Makira, WCS does the managing on behalf of the Malagasy government, and Madagascar National Parks is not involved. Rather than a national park, Makira is a semiprivate reserve that was established as a REDD (reducing emissions from deforestation and forest degradation) pilot project and receives funding by generating carbon credits through avoided deforestation.[51] The park also benefits from a portion of USAID's US$22 million funding provided to Tetra Tech. WCS and its partners have worked to set up dozens of community-based forest management associations in villages bordering the park (also referred to as COBAs or VOIs, as with Marojejy). These groups have been designated to collectively manage 335,173 hectares of land surrounding the core conservation area.

In many ways, Makira has tried to remedy the shortcomings of Masoala. Because a large number of people live within the park boundaries, community involvement is one of the main tenets of the park. WCS establishes contracts to manage land between the communities and the forest administration. Communities sign a three- to six-year contract to manage their land—a contract they themselves are encouraged to devise with the help of WCS employees. The main WCS employee who interacts with the villagers is called the animateur. The animateurs are usually young, educated Malagasy residents of one of the nearby cities of Antalaha or Maroantsetra who travel to their assigned communities for two weeks out of the month in order to evaluate and train the villagers. Each animateur is ideally in charge of three communities, but the animateurs I interviewed were in charge of five.

With the help of the animateur, every month communities execute a field survey, and every six months they write a report to WCS on their activities and performance. Each community's performance is evaluated by the animateur, and each animateur's performance is evaluated by the WCS employee who serves as the animateur's superior. If community evaluations are satisfactory, community contracts are renewed for six years. If they are unsatisfactory, the contract is renewed for only another three years, and elections are typically held to establish new community association (COBA / VOI) leadership.

As with Masoala and Marojejy, community responses to Makira vary. According to one study, nearly half of respondents held a positive attitude toward the park, whereas those more dependent on the forest to generate income remained reluctant and unsupportive.[52] Those who do and do not support the park are likely to fall along the lines of what the animateurs refer to as "good" and "bad" communities. The good communities have fewer transgressions, and the bad communities basically do what they please without regard for park restrictions. There are significantly more bad communities than good. One animateur estimated: "Ninety percent of the people in the communities do not listen to what I say."[53] The especially bad communities receive *pression* (punishment). First, animateurs stop trainings and projects, bringing in the forest chief to issue fines, along with a WCS employee to supervise. For particularly severe transgressions, such as growing marijuana (*rongony*) in the park, an animateur accompanies the forest chief and a gendarme to the village site in order to arrest the offending villagers and send them to court in the nearest city. In general, the fines and taxes of conservation agents and officials who pervade Masoala are not as widespread in Makira. Any government officials who accompany WCS employees, as noted, get paid per diem by WCS for their trips and are specifically monitored to reduce the likelihood of bribes.

Because of its support from carbon credit finance, Makira has more money than Masoala to devote to community development projects. Much of this money goes toward building community offices, which store field reports and evaluations, or toward paying for community patrols. The animateurs also perform "sensibilization"—that is, education and trainings. Animateurs hold trainings on raising rabbits, pigs, and chickens; growing rice; keeping bees; and farming fish and silkworms. They also explain the environmental laws that govern the forest—including provisions of community-level management codes (*dina*)—and help assist with demarcating park limits.

"Makira is strong!" (*matanzaka!*), one animateur assured me as we spoke. The park had already transferred management of 333,100 hectares to eighty-two communities, he noted, in June 2015, with the numbers growing steadily. Only a few villages had refused to participate. Yet he also noted a contradiction. Although delimiting the park was a great success (all of his communities had now marked the park limit, he reported excitedly) and establishing an office for each COBA was also rather successful, most of the other missions of the park remained unfulfilled. Trainings had little impact; many were not appropriate for the local context. A poultry project in one village with no experience raising chickens, for example, resulted in a 70 percent mortality rate due to bacterial disease.[54] Yet trainings continue regardless, and a robust bureaucracy proliferates. Communities continue to file their three-month surveys and continue to receive (often abysmal) evaluations.

Rosewood logging has complicated the sensibilization process. Makira has been subject to similar logging patterns as Masoala, although less intense because the terrain is more rugged and less accessible. From 2009 to 2011 logging in Makira increased, with three hundred to five hundred people transporting an average of three shipments per day.[55] This pattern continued periodically until at least 2015, with operations moving deeper into the forest and requiring a minimum three-day hike into the forest from the closest village. Logs were sent down the Ambanizana River to informal beach ports set up in the Bay of Antongila.

Illegal logging has affected all sectors of Makira, particularly at the park's boundaries, but rosewood logging in particular has been primarily restricted to the eastern sectors of the park. As in Masoala, villagers in Makira often participate in the rosewood trade directly, through logging and trading, or indirectly, by providing services to loggers and traders. During the "hungry period" in March and April just before the rice is ready for harvest, even more people join the trade. Also as in Masoala, but to an even greater extent in Makira, villagers surrounding the park play a special role as "forest owners." Because of their local knowledge, some villagers are hired as guides to help loggers find trees. They are paid 20,000–50,000 ariary (US$5–$13) per day, and if they are considered "forest owners," they earn around 50,000 ariary (US$13) per tree that is found.[56] These wages are of course small compared with the price of the timber as it travels down the supply chain. But they are much larger than anything Makira—or any other park in the region—offers to local villagers. In Makira in particular, owing to the park's rugged and unexplored terrain, the role of forest owner and guide becomes especially important. Yet, at the same time, the question of who actually *owns* the forest—local villagers, global conservation organizations such as WCS, the Malagasy state, or traveling loggers and traders—remains unanswered, as new conservation programs encourage local "participation" yet strive to remain steadfastly in control.

the elusive quest for participation

Northeastern Madagascar provides a prime example of the wider trend toward participatory conservation in the developing world. Although there have been certain successes, conservation in the region has not generally been viewed positively by the people enlisted to carry it out.[57] Rosewood logging in particular has added a new layer complicating conservation efforts. After a precipitous drop in logging following the colonial period, new demand for rosewood in China since the early 2000s has revitalized the hardwood logging economy in northeastern Madagascar. With rosewood far more lucrative than before, loggers and traders are eager to enter protected areas in search of it. Authorities commissioned to curtail the logging instead solicit their slice of the cake, letting logs pass with a fine. Villagers living around the parks in Masoala, Marojejy, and Makira navigate this global junction, deciding at each strategic moment on which side they stand: conservation or logging.

Within this context, the rent-seeking practices of environmental authorities are particularly problematic. These practices ensure that fines will be imposed whether environmental restrictions are obeyed or ignored. They contribute to a worst-case conservation scenario wherein all villagers along rosewood logging trails expect to be penalized regardless of their individual participation in the trade. Whether referred to as the "politics of the belly," "rotten institutions," or "shadow states" or given the terribly oversimplified label of "corruption," this exploitative relationship between authority figures and their subjects is a common stumbling block for conservation efforts.[58] In Masoala, Marojejy, and Makira, conservation managers have enlisted different measures to circumvent the dynamic. Whereas Masoala remains a hotbed for corruption, international conservation organizations in Marojejy and Makira have been somewhat more successful at limiting the interference. They have achieved this by developing close relationships with villagers living in and around the parks and closely monitoring any interactions these villagers have with outside authorities. This has been done, in other words, through a participatory approach.

Local participation provides one way to circumvent the rent-seeking practices of worst-case conservation. But all participatory conservation efforts in the parks of northeastern Madagascar are mediated by international bodies. As much as participatory conservation relies on the presence of local communities, the ultimate goals of the projects are seldom actually shaped by the will of the participants. Rather, the unique innovation of participatory approaches—in Madagascar and likely elsewhere—is that they delimit a space where international bodies can be placed in direct contact with local communities, circumventing to the greatest extent possible the neopatrimonial politics of the state. Despite differences in technique, protectionist and participatory conservation in Madagascar thus retain a core analogy: they both create a fortress. Under participatory approaches to conservation, the problem is not people per se but rather subjectivities that do not prioritize conservation, compounded by an institutional context rife with what is often referred to as corruption.

Figure 4.3. Malagasy rosewood reforested on private land in northeastern Madagascar, approximately five years old (left) and twenty years old (right). © Annah Lake Zhu.

There are other options, however, for pursuing rosewood conservation less emphasized by Western NGOs. Rather than military-style interventions based on criminalization, and rather than molding of environmental subjects that prioritize the protection of nature, another possibility is rosewood reforestation and the creation of nonprotected working landscapes—that is, landscapes that *work,* that are sustainable, if not pristine. This risks the loss of pristine nature, but it likely gains the buy-in of rural Malagasy people, who for the most part look favorably upon the progressive transformation of forest to agriculture. It is, at a minimum, an alternative that any truly participatory approach must consider. WWF employees I spoke with in the region noted they do not grow rosewood in Marojejy because it takes too long to reach maturity. Based on my discussions with interlocutors, this is also the case for WCS in Masoala and Makira. The overwhelming majority of rosewood conservation finance is devoted not to reforestation but to combating the trade, whether via military intervention, helicopter patrols, legal revisions, or new investigatory or judicial bodies and so forth.

Yet some NGOs in the region do engage in rosewood reforestation, notably a local NGO named Macolline, owned by a Chinese Malagasy woman I worked with as a Peace Corps volunteer. Indeed, a few Chinese Malagasy families in the northeast have been pursuing sustainable rosewood alternatives outside the purview of Western conservation. Nearly all rosewood deliberately grown in the region, I was told by a WWF employee, is on these private plantations, where the trees are planted as a type of inheritance for their grandchildren and also in order to sell their saplings locally (Figure 4.3). So far, these are modest parcels—a few dozen hectares here and there—but they might be expanded. As we will see in Chapter 6, China is pursuing such rosewood plantations on a large scale domestically and has also begun to export the model to neighboring Cambodia and Laos. From a participatory perspective, this seems to be a promising route forward. Yet, among the sphere of Western conservation based primarily on the creation of protected areas, it is a wanting substitute for pristine forest.

The goal of participatory conservation, in short, is participation not so much for its own sake but rather to erect a fortress of a different kind. Rather than protecting nature from humanity, participatory conservation protects (or rather *attempts* to protect, given that efforts are often unsuccessful) a transnational space of intervention from the realities of life outside—whether those realities be "corrupt" rent-seeking practices or simply an ethic of living on the land that does not align with conservation. The question of participatory conservation is *how* to preserve resources, certainly not *whether* to preserve them, and in either case the answer is ordained from the outset. Consequently, true participation—if there could be such a thing—remains elusive.[59] Yet if participation (or ownership, or empowerment, and so forth) is in fact the goal, then conservation's most basic assumptions must be put on the table for questioning. The value of pristine nature, biodiversity, and human-free landscapes that so deeply drive conservation efforts must be open to revision. Participation is an issue not of including more voices but of being able to revise one's own assumptions according to what those voices say. It is not participation per se but rather the potential for this fundamental revision—and its power to alter the very question of conservation itself—that remains elusive in the Western conservationist mentality.

5

speculating in species

At the 2019 meeting of the Convention on International Trade in Endangered Species of Wild Fauna and Flora (CITES)—the main international treaty governing the wildlife trade—five southern African nations put forward proposals to relax restrictions on the ivory and rhino horn trade. All of these proposals were voted down, and the Southern African Development Community (SADC, of which Madagascar and the five nations advancing proposals are members) issued an objection. At the conclusion of the meeting, a delegate on behalf of SADC delivered an impassioned speech cautioning against the direction in which the treaty was heading. As currently implemented, the delegate declared, "CITES discards proven, working conservation models in favor of ideologically driven anti-use and anti-trade models. Such models are dictated largely by Western non-State actors who have no experience with, responsibility for, or ownership over wildlife resources." Instead, he continued, these Western actors represent "protectionist NGOs whose ideological position has no basis in science or experience and is not shared in any way by the Member States of SADC and their people." "The time has come," he concluded, "to seriously reconsider whether there are any meaningful benefits from our membership to CITES."[1]

The meeting was reported as one of the most controversial yet—a win for (Western) conservation and a loss for southern Africa, which contains more than three-quarters of the world's elephants and nearly all of the remaining rhinos.[2] Beyond their rejected proposals, African delegates critiqued the voting system more broadly as "tainted, rigged and not free and fair." They pointed to the tendency of richer, conservation-minded actors to buy the votes of poorer nations, tipping the scales in their favor. "It is very disturbing," the delegate observed in his speech, "to see the North / South divide across the African continent rearing its head again." Indeed, this dynamic reproduces at the international level patterns of neocolonial conservation found deep within the forests of Madagascar. As discussed in Chapter 4, no matter how "participatory" conservation might allege to be, endangered resources are *transnationalized* rather than locally owned or managed; they become objects of global concern, and thus the entire international community must weigh in on their conservation. In the process, African lands and resources provide what political ecologist Bram Büscher refers to as an "inverted commons": they are thought to belong to the entire globe, not to the people who rely on them most.[3] Those countries, those people who bear the burden of the conservation, have a small voice on the global diplomatic stage that determines the future of endangered resource management. Occluding this grim reality with the fanciful rhetoric of "participation" and "inclusion" is what journalist John Mbaria and ecologist Mordecai Ogada call "the big conservation lie."[4]

Beyond critiques of eco-colonialism, however, there is another compounding factor when it comes to conventional endangered species conservation efforts. Trade restrictions through CITES—the go-to conservation approach for high-value resources—artificially increase resource scarcity and, consequently, value. Indeed, with rosewood, ivory, and rhino horn, trade restrictions have been known to trigger intensive bouts of speculative investment, ironically having the opposite effect as initially intended. Much to the chagrin of conservationists around the world, endangered species provide a compelling new financial opportunity in the face of waning returns from more conventional investment avenues, and trade restrictions can exacerbate this dynamic. As with Chinese art or antiques, markets for endangered species have become targets of speculation: a way to diversify investment portfolios and hedge against losses in other, more conventional investment avenues. They offer a certain tangibility and cultural prestige that stock markets and real estate lack. Also like art and antiques, endangered species are subject to multiple lines of legality. Their value depends not only on their rarity or cultural appeal but also on their perceived legality. Designation of illegality in one geographic milieu (the West) or scale (the international) is often non-isomorphic with that of another (southern African countries harboring the world's last elephants and rhinos, China and its increasing global reach). Wildlife trade restrictions are not necessarily applied or enforced uniformly across the board. Even worse, they are often perceived as biased or illegitimate, encouraging non-compliance as a form of protest. As clearly demonstrated in the SADC representative's impassioned speech, international law is seen as the product of Western conservation ideologues throwing their weight around in the global arena. There is little incentive to comply.

This chapter explores the complexities of restricting the trade in endangered species given its speculative dynamics and the legal and cultural relativism it brings up. Because of financial speculation compounded by conflicting lines of legality, international restrictions have unintended consequences, both on the demand side in China and on the supply side where species are sourced. Within China, international trade restrictions can trigger market booms, resulting in what Chinese timber traders have come to call, as noted previously, "the Madagascar phenomenon" (*Mădájiāsījiā xiànxiàng*).[5] Outside of China, on the supply side, international trade restrictions have further unintended consequences. The illicit status of endangered species typically reinforces local inequalities, allowing only elite figures with clandestine connections to gain access to these markets (as discussed in Chapter 3). The fight against the trade—now made illegal—becomes a military operation, often co-opted by the same elite who are involved in trafficking. The "war on poaching" is not unlike the "war on drugs."[6] Criminalization empowers an elite minority and stigmatizes entire communities. Yet, as aware as we are of this dynamic for narcotics, we often overlook the analogy with the "war" against trafficking of endangered species.

Beyond demonstrating the inadvertent impacts of restricting the trade in endangered species, this chapter also examines the highly controversial question of trade legalization and captive breeding. As an alternative approach to criminalization, captive breeding garners strong support from certain circles in China and many other countries with endangered species, but it receives fierce opposition internationally. In light of the global COVID-19 pandemic—likely connected to a wet market in Wuhan, where both wild-caught and captive-bred animals were sold live—the controversy over captive breeding has greatly intensified. Many in the United States and Europe advocate for a total ban on wildlife trade and consumption, including most captive breeding operations. China has responded swiftly, banning all terrestrial wildlife trade for consumption and making substantial revisions to the country's Wildlife Protection Law regulating the industry.[7] But the ban retains exceptions for wildlife used in medicine, as well as continued domestic support for captive breeding programs to save endangered species.

In general, the issue of captive breeding, especially as it relates to wildlife consumption and traditional Chinese medicine, has become an issue in which neither side is willing to give much ground. Similar to the polarized issue of rhino and elephant conservation in southern Africa, it incites strong and opposing responses that are becoming more entrenched over time. Moving past ideological rallying cries from both sides, the goal of this chapter is not to adjudicate between these different positions or offer resolution but rather to move toward a more complete understanding of the diverse forces that animate the debate.

international trade restrictions trigger speculation

Since being established in 1963 and entering into force in 1975, CITES remains the primary means of conserving endangered species by restricting the international wildlife trade. Parties to this convention (the countries that have signed and ratified it: 183 in total, indicating near universal participation) meet every two to three years to list (or in some cases de-list) additional species—that is, to prohibit their trade to various degrees. Many of the species listed in CITES—elephants, rhinoceroses, tigers, pangolins, and certain species of rosewood, ginseng, sharks, turtles, and tortoises—also have speculative appeal in China. In some of these cases, trade restrictions and stockpile destructions commonly recommended under CITES are not as effective as planned and can even backfire, triggering speculation.

Ivory and rhino horn provide classic examples. These are what the United Nations Office on Drugs and Crime (UNODC) refers to as "investment-grade wildlife products."[8] Trade restrictions or stockpile destructions can end up increasing the value of these resources, initiating a vicious cycle, as UNODC explains: "Unlike cocaine or heroin, there is an absolute limit on the amount of ivory that can be produced, so there is a danger of a vicious cycle ensuing, where each elephant poached increases scarcity, and thus the incentives for poaching another. Paradoxically, interdiction and destruction of ivory stocks would also serve to limit supply, further enriching those invested in ivory."[9]

In other words, stockpile destructions reduce global supply, "concentrating ivory market power with speculators illegally [or legally] holding ivory . . . exerting upward pressure on prices and making extinction trajectories more likely."[10] Some speculators have maintained ivory stockpiles for more than three decades since the initial CITES listing. Ivory dealers in Hong Kong, for example, continue to maintain more than one hundred metric tons of stockpiled ivory since the CITES listing in 1990.[11] Destruction of confiscated stockpiles makes illicit stockpiles that have not been confiscated much more valuable. It also makes the economic incentive to harvest the last the market has to offer—the last remaining wild populations, that is—that much greater.

In addition to ivory, international trade restrictions for rhino horn similarly contribute to hoarding and speculation. Worth as much as US$65,000 per kilogram at its peak—more than gold, diamonds, or cocaine by unit volume—rhino horn is a ripe investment opportunity in China. Demand for rhino horn, as with ivory, is inelastic, meaning it does not decrease as price increases. In fact, it is likely to do the opposite, *increasing* with price increases that demonstrate the investment quality of the resource. As rhino horn has become increasingly rare, its price has grown higher, pushing the resources into what has been referred to as "an economic supply-and-demand extinction vortex."[12] Within this context, CITES "artificially restricts supply in the face of persistent and growing demand," in the end inadvertently bolstering the supply-and-demand extinction vortex.[13] It does not reduce trade but rather contributes to the buildup of stockpiles across Asia and Africa as speculative investments. Moreover, CITES decisions do not consider the impact that proposed restrictions would likely have on consumer demand and prices, making the process impractical and the recommendations not grounded in economic realities.[14]

The unintended consequences of CITES listings have been most clear for rosewood. The wood's speculation has become so extreme in China that prices of the most endangered varieties are now directly manipulated by a small group of professionals. In recent coverage, China Central Television described this phenomenon as the "rosewood industry speculation" chain, in which certain speculators buy up a substantial supply of a given species and hoard it for a year or longer to trigger its scarcity. As soon as the price spikes, the speculators channel the wood through various first-, second-, and third-level dealers, who all work together to raise the value as much as ten times higher than the original purchase price. Denouncing this process, Chinese media outlets characterize rosewood speculation as "even worse than robbing a bank," triggering a volatility that can be "more ferocious than real estate."[15] In addition to these elite investors, there are a variety of amateurs—up-and-coming entrepreneurs with more money in their pockets than ever before. Because it is harder to distinguish between licit and illicit rosewood varieties (whereas *all* elephant ivory and rhino horn is illicit), a broad range of investors, from middle-class families to billionaire industrialists, feel comfortable entering into this lucrative market.

Rather than stopping the trade in high-value species such as rosewood and ivory, international trade restrictions often drive it underground. Harvesting becomes poaching; trade becomes trafficking. This can have devastating effects on the supply side, where entire economies are driven underground with little to no benefit to home countries. Instead, the trade continues through clandestine circuits. Ivory and rhino horn trafficking, for example, are said to be "run by the same criminal syndicates" as those trafficking narcotics.[16] These criminal groups are further alleged to have ties to larger terrorist organizations, leading to the claim that ivory is the "white gold" of jihad, along with the wider accusation that wildlife trafficking poses a threat to global security.[17] Whether or not this is the case, traffickers of high-value endangered species are indeed an elite and well-connected few operating through illicit networks. Their profits, whether used for terrorism or siphoned to overseas accounts, are for the most part not repatriated to the countries harboring the species and generally go untaxed and unaccounted for. As observed in the CITES proposal put forth by Eswatini (formerly Swaziland), "at present 100% of the proceeds from the sale of rhino horn are taken by criminals, while rhino custodians pay 100% of the costs of rhino protection and production without the funding that could cover these costs from legal trade."[18]

In terms of criminalization, elite traffickers often remain immune to indictments stemming from the trade. Rather, it is loggers and poachers at the lowest levels who bear the brunt of the charges and communities living closest to endangered species that feel the criminalization most. Given the extreme prices, working in the trade in endangered species even at the lowest levels is quite lucrative compared with other economic activities and often considered well worth the risk. With already tense relationships between conservation areas and local communities feeling expropriated from their land and resources (see Chapter 4), there is also typically not much of a moral stigma for those working in the trade. Indeed, many even consider their work to be not illegal but within a zone of "contested illegality."[19] The critique made by southern African countries at the 2019 CITES meeting, along with their threatened withdrawal from the convention, clearly illustrates this contested legality. On the ground, in African countries where the wildlife trade can be highly lucrative, trade restrictions are contested more deeply. As my interlocuters have questioned me concerning rosewood, why leave a tree that is worth US$60,000 per ton in timber markets in China standing thousands of miles away in the forests of Madagascar? Wildlife conflict provokes more extreme questions. Why protect an elephant that destroys harvests and occasionally targets people as well? Elephants kill an estimated five hundred people per year, in addition to causing extreme crop damage.[20] Often it is not "poachers" who kill elephants but hunters and locals threatened by the elephants' presence.[21] Why not let the people threatened by their encroachment have a greater say in how the elephants are managed?

Combating the trade with sanctions does not ease this tension but instead has resulted in the "militarization of conservation," wherein conservation objectives are executed in a militaristic fashion by state or private security forces.[22]

The poachers, too, are well armed, and this results in a "conservation-related arms race" to poach or protect endangered species.[23] The dynamic benefits very few—and there has yet to be compelling evidence that it protects the endangered species.[24]

International trade restrictions, thus, have two major unintended effects: (1) artificially increasing price on the demand side while (2) driving the trade underground and encouraging militarized conservation on the supply side. The result is thriving speculative markets for endangered species alongside an increasingly militarized supply chain. With the classic examples of ivory and rhino horn, this has been true for decades. International trade restrictions for these species were set so long ago—primarily in the 1970s, 1980s, and 1990s—that it is difficult to demonstrate their precise impact. During this time, China was only beginning to experiment with a market economy, and the speculative dynamics of today's markets for endangered species were still far off. Only since 2005 has market speculation for Chinese cultural artifacts and materials—including endangered species—escalated drastically. Rosewood from Madagascar, for example, was worth next to nothing in China in the 1980s and 1990s, but it was valued at US$60,000 per ton by 2013. Similarly, illicit ivory prices in China from 2006 to 2013 ballooned from around 5,000 to 20,000 RMB per kilogram, matched by an increase in poaching during the same time.[25] For rhino horn, the trend is even more extreme, with the retail price per kilogram increasing from US$5,000 per kilogram in 2009 to US$97,000 in 2014, with the average annual number of rhinos poached increasing from 122 to 1,215 over the same period.[26]

Outside of the iconic species, however, other recently listed species demonstrate how international regulations can more directly trigger price increases. Rosewood provides the clearest example. International trade restrictions in rosewood illustrate the tight connection between CITES listings and market speculation within the past decade. In 2013, a large number of endangered rosewoods were added to the convention. The new trade restrictions contributed to unprecedented price spikes in the listed species, making rosewood a ripe buying opportunity in China. "The increased enthusiasm for rosewood," the chairman of a classical furniture association explained in reference to CITES, "has come mostly because of the release of new international regulations."[27] This was so clear for the listings in 2013 that it was deemed the "Madagascar phenomenon." Subsequent CITES meetings in 2016 and 2019, during which additional rosewood species were listed, also proved to stimulate the market, although not to the same extent as in 2013, when a larger number of more commonly traded species were listed. These speculative jolts can cause increased logging on the ground. In addition to increased logging in Madagascar, a recent study found that the exploitation of rosewood from continental Africa (*Pterocarpus erinaceus*) increased by 129 percent, and the incidence of illegal trade increased by 120 percent, after the CITES listing in 2016.[28]

The speculative jolts that CITES delivered to rosewood markets is not evidence of the convention's ineffectiveness. To the contrary, the fact that CITES regulations trigger rushes to buy while supplies last and the massive stockpiling of listed resources demonstrates their—at least partial—efficacy. The problem, however, is that even when banned or restricted, species maintain high values. They often *increase* in value, although only in those geographic milieus where restrictions are less likely to be upheld. For those who can acquire them, endangered species remain an excellent investment opportunity—an even better investment opportunity, in some cases, than if they were freely traded. On top of this, the chance of extinction makes these resources even more valuable and their speculative potential that much greater. If trade restrictions are strictly enforced *domestically*, however, the dynamic changes.

domestic restrictions

Unlike international trade restrictions, domestic restrictions—especially those instituted in the countries where demand is highest and that are backed by popular support—can have a beneficial effect. This has been the case with certain regulations promulgated in China. One of the most notable examples is an anti-corruption campaign that was, ironically, not intended to address the wildlife trade at all. After taking office in 2013, President Xi Jinping embarked on a massive anti-corruption campaign that hit many luxury markets hard—including those for endangered species. Although the goal of the campaign was to crack down on bribery and favoritism within the Chinese Communist Party, the result was a dramatic decline in sales of endangered species such as rosewood. Afraid of being targeted by the campaign, business interests courting local political figures stopped offering spacious condos replete with rosewood furnishings as a token of their support. Cognizant of this political overhaul, investors seeking the highest returns thought twice about putting their money into rosewood.[29] Xi's anti-corruption campaign has reduced sales of other endangered species as well. Large banquets serving shark fin soup to prominent politicians were canceled and expensive medicinal concoctions gifted in the spirit of seeking favors were reconsidered, all because of this crackdown on corrupt consumption.

Xi's anti-corruption campaign has arguably done far more to stifle the trade in endangered species in China than any internationally imposed trade or logging restrictions. Even though these impacts have been for the most part accidental (the campaign was certainly not intended as a conservation measure), other domestic regulations have been established specifically to protect endangered species. China's ivory ban in 2017 has become an iconic example of this. Unlike international restrictions, Chinese domestic prohibitions send strong signals to investors of weakening future potential. As long as domestic prohibitions are considered permanent, are fully enforced, and are accompanied by reliable penalties—which seems to be the case with China's ivory ban thus far—the speculative potential of the resource weakens. Investors offload stocks in anticipation of the prohibitions, potentially leading to a "fire sale," with prices plummeting and speculative investment severely disincentivized.[30] Thus, domestic prohibitions, if implemented properly, have the opposite effect of international restrictions that are not fully implemented or enforced in home countries.

China's ivory ban has thus far been considered a success in reducing demand and speculative appeal. The price of ivory both within China and in range countries has declined substantially, and market turnover has decreased, since the ban's implementation.[31] Yet with rhino horn and tiger bone, China has chosen a different route. Officially, trade in rhino horn and tiger bone was banned in China in 1993 to comply with CITES. These regulations were in many ways instituted begrudgingly, and since this time China has been flirting with the possibility of legalizing the trade. In fact, in 2018, the State Council of the People's Republic of China issued a policy lifting the ban on the rhino horn and tiger bone trade under certain conditions. Although the policy strengthened the ban when it came to trafficking parts sourced from wild animals, instituting further measures to crack down on this illicit trade, it simultaneously permitted a legal trade for the medicinal use of captively bred tiger bone and rhino horn. After only two weeks and a dramatic backlash from the international community, China's State Council reinstated the previous ban, indicating that implementation of the new policy lifting the ban would be delayed until an unspecified future time.

All of this begs the question: Why, having pioneered a comprehensive, well-enforced domestic ivory ban, does China remain hesitant to do the same with rhino horn and tiger parts—and even more so, rosewood? And how might China be convinced to implement permanent and well-enforced bans for endangered resources beyond ivory? Answering these questions requires an examination of the long-standing, highly divisive issue of captive breeding.

"Tiger Kings" and tiger bones

The legacy of captive breeding in China differs drastically from that of the West. Consider the case of entrepreneur and animal breeder Zhou Weisen, who has come to be known as China's "Tiger King"—not to be confused with the American tiger breeder / borderline cult leader "Joe Exotic," now popularized by the Netflix documentary of the same name. Born into rural poverty in 1962 (the Year of the Tiger in the Chinese Zodiac), Zhou made a name for himself in his local community at the age of sixteen as a duck breeder. In the 1980s, under Deng Xiaoping's regime of economic liberalization, Zhou made his fortune breeding and exporting snakes to collectors in Hong Kong and Macao. Although Zhou had only an elementary school education, his ultimate ambition was to be the first person in China to breed and display tigers. In 1993, with approval and financing from China's State Forestry Administration, Zhou opened the Xiongsen Bear and Tiger Villa, a resort complex that combined breeding facilities with zoo attractions, hotels, and restaurants. In just a few years, Zhou's villa grew from housing 60 to more than 1,500 tigers, becoming a premier tourist attraction in the Guilin region.

Unlike the American "Tiger King," Zhou is widely represented in China as a respectable entrepreneur and animal lover: one among many rags to riches successes in post-Mao China. "I am trying to realize the value of life," he said to one reporter in 2014. "If people all over the world know that a man named Zhou Weisen has saved the tiger that was once extinct, I feel that my life value will be realized."[32] In the West, however, Zhou and his villa have received unanimously negative media scrutiny. He is often labeled a "tycoon" or "millionaire," and his tiger breeding compound has been described as "shabby," "dispiriting," and "squalid"—a nightmarish "circus" where tigers are industrially manufactured only to be harvested for wines and other products of questionable to no medicinal value. These representations paint Zhou as a toned-down Chinese version of the American "Tiger King," an outlandish criminal profiting from the illicit breeding of endangered species.

The disconnect between Chinese and Western representations of Zhou and his breeding business—as well as tiger breeding more generally—represents a larger divide in an ongoing debate between China and the West over the issue of endangered species and captive breeding. It is very similar to the debate over rhino horn in southern Africa, where large ranches breed rhinos and trim their horns with little harm to the animal, as if administering a nail clipping. Both tiger farmers in China and rhino ranchers in southern Africa maintain stores of bones and horn clippings, respectively, that cannot be sold because of CITES rulings. The tiger / rhino debate has multiple layers with compelling arguments on both sides. More than just a question of whether or not captive breeding can save a species, however, the debate centers on a deeper divide over what exactly it means to "save" a species.

China has a long history of supporting captive breeding initiatives—such as Zhou's—as a way to supplement or restore wild populations and meet domestic demand for these species. The country's first Wildlife Protection Law in 1988 encouraged captive breeding of various endangered species, including tigers, bears, and pangolins, through both state-sponsored breeding operations and small-scale backyard breeding by individuals across the country. When China banned the trade in rhino horn and tiger parts in 1993 to comply with CITES, the government also instituted a push for the development of captive tiger breeding in hopes that a legal trade in captive-bred parts could eventually serve as a source for consumer markets. Practitioners in China were also well aware of rhinoceros breeding in southern Africa and the potential to develop a legal market there as well. To Chinese practitioners, this dual strategy made perfect sense: ban the wild-caught trade while developing captive breeding capacity.

Such a strategy, the Chinese government has found, could address other needs as well. Beyond the market demand for animal parts and the environmental concerns for the future of endangered species, there is the state's desire to develop businesses in rural regions. Breeding sites are thought to offer economic benefits for rural areas through direct sales of animals or, in the case of larger charismatic species, tourism. Captive breeding of tigers and rhinos in particular is also compelling because of their strategic importance in traditional Chinese medicine and the soft power potential this form of medicine serves for a globalizing China. Traditional Chinese medicine, as noted in Chapter 1, is not only a set of medical practices but a cultural heritage and bedrock of Chinese identity. It figures prominently in China's plans for increasing its sphere of global influence, on not only political and economic but also cultural fronts. Thus, international efforts to curtail the trade in these species are met with well-entrenched resistance. Western-style conservation, with its antagonism toward captive breeding and its call to ban all consumer goods made from endangered species, is received as a challenge to China's cultural sovereignty domestically and its ambitions for cultural expansion abroad.

Thus, given that there is no definitive evidence that captive breeding operations threaten wild populations on one hand, and given that there remain prospects of meeting future demand through captive-bred populations on the other hand—in other words, given the multiple forces pushing for a legalized trade and no categorical evidence demonstrating its negative effects on wild populations—the Chinese government is very hesitant to retain a comprehensive ban. This is reflected in the country's initial 1988 Wildlife Protection Law as well as the 2016 revision—both of which support captive breeding operations, given that they are strictly permitted, monitored, and regulated, with specific requirements to ensure animal safety and welfare (Articles 25, 26, 28, and 29 of the 2016 Wildlife Protection Law). The 2021 revision has changed aspects of the law in light of the COVID-19 pandemic, more than doubling the number of protected species. But captive breeding as a conservation strategy remains codified in the law.

This is especially true for the captive breeding of tigers, the most controversial of China's captively bred species. Breeding tigers has been ongoing for decades in China. Concurrent with the 1993 ban on selling tiger bone and rhino horn, the State Forestry Administration allocated funding for two tiger breeding operations: Zhou Weisen's Xiongsen Bear and Tiger Villa in southern China and the Feline Breeding Center in the northeast. By 2007, there were 5,000 captive tigers in China and no more than 50 wild tigers.[33] Today, the country holds roughly 6,000 captive tigers with still no more than 50 wild tigers (Figure 5.1). It has been estimated that captive tiger populations might reach 70,000 to 100,000, given proper support, but because of the controversy surrounding the issue, captive tiger populations remain stable. The State Forestry Administration is planning various rewilding schemes in northern China, with hopes of expanding to

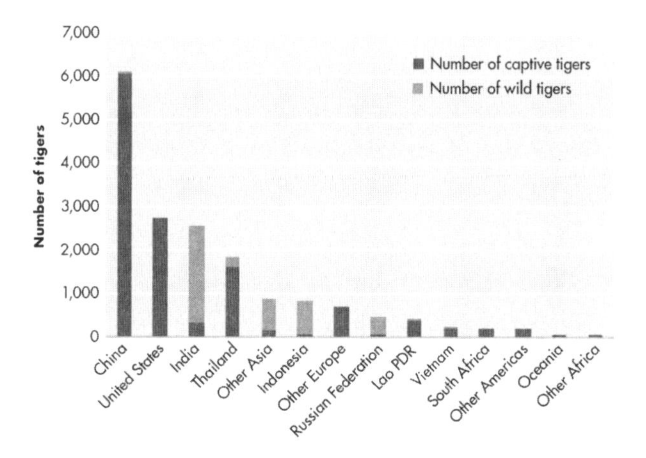

Figure 5.1. Number of wild and captive tigers by country, based on 2016 or most recent data. United Nations Office on Drugs and Crime (UNODC), *World Wildlife Crime Report: Trafficking in Protected Species*, 2020, Figure 9.

southern China as well. Most of these proposals, however, have been strongly opposed by the international conservation community.

international opposition

The World Wildlife Fund (WWF), TRAFFIC, the Environmental Investigation Agency, and other mainstream conservation groups contest China's position on captive breeding. Their arguments against the practice are as simple and compelling as the arguments for it: legalization—whether for captive-bred or wild individuals—stimulates the trade. It is prohibitively difficult to discern between products sourced from captivity or the wild, detrimentally blurring the boundaries of legality in a way that ultimately favors poaching. Moreover, legalization reduces the stigma (which admittedly may not have been there in the first place) of consuming endangered species and makes them more readily available for purchase, encouraging additional users to take part in the buying.

The case of the 2008 one-off selling of confiscated ivory stocks provides the go-to example of how legalized sale can backfire. After this onetime sale, demand for ivory increased dramatically. The timing of the sale, however, was unfortunate for the sake of a reliable counterfactual. Demand for ivory had already been increasing, and the global financial crisis made investments in endangered species and other cultural goods more expedient than more conventional investment avenues.[34] Consequently, causality has been difficult to establish. Another go-to example is the case of bears captively bred for bile, a prominent ingredient in traditional Chinese medicine. Since legalization of the captive breeding and trade of bears for their bile, demand has increased, with an array of products related to bear bile now available on the market. The same fate, many conservationists fear, is likely to await the tiger if that trade is legalized as well.

But if you consider a different animal, a different outcome is also possible. Indeed, there are many highly successful captive breeding programs operating in the United States. Despite facing near extinction in the mid-twentieth century, for example, American alligator populations have undergone a "complete recovery" and are now considered a species of least concern by the International Union for Conservation of Nature. Captive breeding—both for harvest of skins and for reintroduction into the wild—contributed significantly to this recovery. Alligator farming is a multimillion-dollar industry in the United States, with more than ninety licensed alligator farms generating US$7.6 million in the state of Florida alone.[35] Although the American alligator is still listed in CITES

Appendix II, the legal sale of captive-bred alligator hides not only is permitted but has undeniably contributed to the species' survival.

Captive breeding of both crocodiles and alligators is largely considered beneficial for conservation—at least as it occurs in a Western context (the United States and Australia in particular). Subsidized by tourism revenues, captive breeding operations collect eggs from the wild and also produce them from their captive stock. They typically return a certain percentage of juveniles to the wild after being reared. Purely conservation-based operations return them all. Beyond crocodiles and alligators, this type of captive breeding to meet market demand is being initiated for other rare exotic pets, plants, reptiles, and seahorses.[36] For many of these species, captive breeding has been shown to have an overall benefit for wild populations as well as to provide livelihoods for local communities.[37]

For tigers, rhinos, and other charismatic megafauna, however, there is sharp disagreement, with most range states in support of captive breeding and a legalized trade and nonrange states and international organizations largely in opposition.[38] Rewilding of these large mammals is of course more difficult than for species that do not require such large ranges, are not raised directly by their parents, or are less likely to cause human / wildlife conflict, further contributing to the disagreement. The debate has been ongoing since the 1990s, with little progress made on either side in determining the best way forward. "The reason that this debate generates more heat than light," Save the Rhino rightly observes, "may simply be that no one can know for certain what the real impact of opening up such a trade would be, and there are plausible arguments that it would be positive or negative." "Beware," the organization continues, "anyone who makes definitive statements about an uncertain future."[39] This observation cuts to the heart of the issue and extends far beyond the trade in rhino horn.

For many endangered species—including rosewood—it is unclear what exactly the consequences of either trade legalization or criminalization will be. Part of the difficulty of comparing markets for endangered species is the subtle difference in how each is commodified and consumed. As commodities, tiger bone and rhino horn fall somewhere in between art objects for display in the home (such as ivory or rosewood) and prized medicinal goods to be prescribed and consumed regularly (such as ginseng or bear bile). Because of their high cost, they resemble the former, largely purchased as elite cultural symbols, gifts, or speculative investments. Yet, because of their purported medicinal value, they also resemble the latter, used as a preventive measure against disease or aging (although rhino horn is typically used for only acute medical conditions). The market for goods such as ivory and rosewood makes little to no distinction between wild and farmed products, whereas the market for ginseng and bear bile does make a claim of greater efficacy for wild over farmed. How exactly legalization or criminalization of the sale of captive-bred tiger and rhino parts will impact consumer habits will ultimately determine how these different options will impact demand. Market speculation, of course, adds another compounding factor, making species more valuable the more endangered they are. For species that are the target of speculation, increasing availability through captive breeding and legalization is likely to decrease speculative appeal, though it may also simultaneously increase nonspeculative demand.

Internationally, the COVID-19 pandemic has bolstered the long-held opposition to China's captive breeding efforts. This pandemic has made wildlife trade and consumption an issue that concerns everyone around the world. Formerly obscure practices such as wet markets and commodities such as pangolin scales and bear bile have gained international notoriety. Along with that attention has come increasing politicization and ideological polarization. Environmentalists point to the outbreak as a clear demonstration of the unintended consequences of the wildlife trade, arguing for its complete cessation. China has, to a large extent, followed through with these requests, banning all farming and consumption of terrestrial wildlife within the country in February 2020, with subsequent white-listing of certain species. The Wildlife Protection Law has undergone a number of amendments, but there remain provisions for the trade in wild animals for medicine, pets, and research. It seems unlikely, in other words, that China will shut down its captive breeding programs in tigers anytime soon. The goal appears to be stringent oversight and regulation of all captive breeding operations that are allowed to continue moving forward, such that they do not become breeding grounds for viruses or illicit trade as well. Underregulated and overcrowded urban wet markets have a much greater potential for viral mutation than state-sponsored breeding operations, given proper oversight.

the deeper divide

Beyond these well-trodden arguments and even the more recent developments surrounding COVID-19, there is a deeper divide between Western, Chinese, and southern African approaches to wildlife in general that makes common ground evasive. Even if captively bred tigers might meet demand for their parts without threatening a global pandemic, it would ultimately be beside the point. From a biodiversity conservation perspective—and more generally from a perspective that privileges wild over domestic spaces—there is little value attached to captive-bred individuals that have no possibility for reintroduction into the wild. Captive populations simply have no conservation value as long as they stay captive. Beyond that, even those individuals that might be reintroduced pose the threat of "genetic pollution," the dilution or alteration of wild-type genomes caused by the introduction of those from captive-bred gene lines.

Biodiversity conservation depends on maintaining distinct species within their natural home range. There might be one hundred times as many captive individuals as wild, as is the case with tigers in China (or millions more, as is the case with the endangered Chinese salamander); nonetheless, the former is infinitesimally less valuable than the latter from a biodiversity conservation perspective. Fighting *extinction* is not the goal; rather, the goal is to fight *extinction in the wild.* Without the potential for a viable reintroduction that does not compromise natural gene lines, captive-bred individuals cease to have any conservation value.

This differential valuation of wild and captive-bred individuals exists for rosewood as well. It at least partially explains the persistent focus on the part of Western nongovernmental organizations (NGOs) on stopping rosewood logging rather than increasing rosewood planting or sustainable forestry more generally. The ultimate objective of the major environmental NGOs working in Madagascar—Conservation International, WWF, the Wildlife Conservation Society—is to protect the island's last wild spaces. All along its eastern seaboard, Madagascar has lost its primary forests to agriculture and secondary growth, while northeastern Madagascar is home to some of the largest and last remaining tracts of contiguous tropical forest. The NGOs operating in the region see it as their mission to protect these primary forests. The loss of rosewood from the forests is symbolic of the loss of the last wild spaces in the region. Establishing rosewood plantations—while potentially reducing erosion, increasing water quality, sequestering carbon, and generating livelihoods—would do nothing to save these remaining tropical forests. Consequently, the ultimate goals of Malagasy people and Western conservation organizations operating in northeastern Madagascar are by and large not the same.

Along these lines, but going beyond the science of conservation, there is also a general sentiment among Western conservation organizations concerning the sanctity of wildlife that diverges from how wildlife is commonly viewed in non-Western contexts, including Madagascar and China. In the United States and Europe, there is the desire to protect the integrity of a majestic tree such as rosewood or animal such as the tiger and save it from the fate of being merely a plantation crop or domesticated farm animal. This contrasts with, but is ultimately analogous to, China's own concerns about protecting the integrity and value of traditional Chinese medicine. Western audiences point to the audacity of conflating the dignity of a wild tiger with mere domestic "herds"—reducing these apex predators to the lowly chicken, pig, or cow.[40] They would likely say the same regarding rosewood: a plantation is fine, but it is no replacement for the forest. Deeper than the desire to save rosewood and tigers from extinction, and deeper than the desire to conserve the genetic diversity of the species, lies the power of these species as untamed symbols of the wild. Against the enchantment of a five-hundred-year-old tree slowly maturing in the forest, digging deeper by root and extending farther by branch over centuries, a human-made plantation fated for eventual harvest pales in comparison, no matter how threatened the trees growing within it may be. The contrast is even more severe with tigers. Against the romance of the tiger freely roaming in its wild habitat, the notion of a tiger farm, with its concrete enclosures, sterile breeding labs, processing facilities, and freezers and tanks full of tiger parts, is repugnant to conservationists, as well as to most Westerners in general.

In China, however, these species have a different allure. Earlier chapters in this book have detailed rosewood's allure as a symbol not of pristine nature but rather of the productive unity of nature and culture. Human ingenuity and intention combine with the beauty of a natural substance in a profound synthesis that represents the height of Chinese aesthetic expression. Concerning tigers, one can observe a similar dynamic. As a profoundly cultural symbol, the tiger represents not pristine nature but a long tradition of the intermingling of Chinese culture and environment, demonstrated not only through millennia of traditional Chinese medicine but also through an equally long history of folklore concerning the species.

Tigers and people intermingled in what is today's southern China long before the earliest Han frontier settlements in the region. The Han Chinese approach to tigers—and perhaps even the Mandarin word currently used, *hǔ*—is believed to be built on these indigenous mythologies. Analysis of the folklore of indigenous tribes reveals the deep cosmological significance attributed to these fearsome predators, including widespread tales of humans transforming into tigers after their death. As Han settlers arrived and settlements grew, tigers and people were pushed closer together and attacks became increasingly common, occasionally with villagers killed in their own homes as tigers pushed down doors and entered houses. One of the longest-standing records of human / wildlife conflict—county gazetteer records across four Chinese provinces extending from 48 to 2000 CE—demonstrates this increase in tiger attacks beginning in the fourteenth century.[41]

Despite increasing attacks, tigers were not viewed as persistent menaces to exterminate without apology, as with bears and wolves in the Euro-American context. "Though near-extermination has been the final outcome in the last phases of the ancient and on-going human-tiger relationship in China," writes ethnographer Chris Coggins, "the cultural significance, especially the high aesthetic, totemic, spiritual, and medicinal value of the tiger in China, has had no parallel in the belief systems of Christian Europe of Euro-America."[42] Coggins writes of the sixteenth-century *Tiger Compendium* (*Hǔhuì*), a collection of tiger lore from centuries prior revealing that these majestic beasts were both feared and revered, were thought to be rational much like people, and were occasionally even taken to court and tried for their crimes against humanity. Unlike other magnificent predators that dominate Judeo-Christian culture, the tiger in Chinese culture was thought in certain instances to be able to serve the will of God by enforcing celestial laws. Moreover, the tiger's medicinal power—both in the flesh through its bones and body parts and symbolically through hats and other objects with tiger-face motifs—has no parallel in Western culture.[43]

As a result of this multimillennial cultural legacy, and as a result of larger differences between cultural valuations of nature, breeding tigers in captivity does not pose the same immediate aversion in China as it does for most Western observers. The tiger is a powerful symbol in China—not of untamed nature but rather of Chinese culture. Tiger preservation is not only a scientific issue but also a cultural ideal, demonstrating the historical continuation of a people within a wider landscape. The continuation of the tiger in Chinese landscapes—wild or domesticated—is the goal, and there is no insurmountable difference between these wild and domesticated spaces. Thus, the question of captive breeding and a legalized trade is not just a question of whether or not these strategies can save the tiger (or rhino and so forth) but of *what sort of tiger* (and rhino and so forth) *is worth saving.*

In Chinese, there is a slogan in support of tiger breeding programs: "Saving tigers with tigers" (*yǐ hǔ yǎng hǔ*). This refers to the idea of saving the species by breeding it; that is, the tiger can be saved by its use as a tourist attraction or through the trade in its products: bones, organs, and fur, all of which will provide funding for tiger rehabilitation and habitat protection. From a Western perspective, however, the slogan can just as easily be understood as a call to save the species with its abstraction. Whether through its ecological role as an apex predator, its genetic value to the planet's biodiversity, or its rich symbolism in Western imaginaries of untamed wilderness, the tiger in the abstract provides a vital charge for protection. The difference in these two readings, as well as the debate over captive breeding and trade legalization more generally, pivots around the word "tiger": its meaning and connotation are in fact quite distinct depending on one's vantage point. Answering the question of what the tiger really is and what sorts of tigers are worth saving pulls into its orbit larger philosophical issues that remain largely obscured in the current debates over population figures and poaching statistics. Questions concerning the value of wild versus domesticated spaces, use versus protection or preservation, turn out to be more philosophical than scientific. This much deeper layer to the debate reveals a much deeper source of deadlock that neither side is particularly inclined to acknowledge.

One thing remains clear from the past decades' impasse: imbricated within the captive breeding debate are a host of other issues surrounding culture, history, race, aesthetics, animal welfare, and ideas of wilderness that all inform different approaches to endangered species conservation. Just as with the rhino horn goblet "stolen" from Drottningholm Palace a decade ago (see the introduction), the question of captive breeding elicits two entirely distinct cultural, geopolitical, and politico-juridical reference frames. From one vantage point, a legalized trade in captive-bred wildlife (at best) blurs the legal boundaries of a trade that should be criminalized and (at worst) serves as a poorly crafted guise for elite financial enrichment at the expense of animal cruelty. From another vantage point, a legalized trade in captive-bred wildlife (at best) paves the way for meeting existing demand for endangered species while repopulating the wild and maintaining national sovereignty and (at worst) does little to affect, positively or negatively, the ongoing slaughter of wild populations by illicit networks. On both sides, there is compelling logic and admittedly no definitive evidence. But also on both sides are conflicting views of nature and culture, deep-seated suspicions of ulterior motives, and a general aversion to fully understanding the positionality of the other. Just as with the different lines drawn between the licit and illicit in a Chinese and Western context, there are different boundaries between captive and wild, leading to different conclusions concerning the question of captive breeding.

rethinking conservation

Endangered species conservation has a noble legacy in the United States and Europe. Indeed, protecting biodiversity, saving wildlife, and marveling at nature seem to be among the few objectively "good" things to which one can aspire in our postmodern, post-truth society. Yet, as this and other chapters have shown, even this pursuit of endangered species conservation—perhaps *especially* this pursuit of endangered species conservation—is plagued by a certain relativity, revealing different approaches depending on one's geopolitical, politico-juridical, and cultural-economic reference. As with pillaged antiques and (neo)colonial appropriations, the debate over the trade in endangered species pulls into orbit a host of other cultural, historical, and philosophical factors that on the surface seem to have little to do with the science of conservation.

Scrutinizing our commonly held assumptions about markets for endangered species allows us to reveal these deeper layers. It allows us to see that, first, demand for many of the iconic high-value endangered resources, such as ivory, rhino horn, and tiger bones, is not solely consumer driven but highly speculative. These resources retain such extreme values because their ancient cultural history has been revived within the context of a highly speculative late capitalist economy. Endangered species in China provide the perfect intersection between cultural allure and rampant financial speculation, promising scarcity-based value in an increasingly volatile global economy that casts doubt on more conventional investment avenues.

Within the context of this rampant speculation, and within the context of growing sentiment against what is often understood as Western neocolonialism, conventional approaches to regulating the trade in endangered species can backfire. International trade regulations reinforce existing scarcities, reaffirming their promise as a ripe investment opportunity that will appreciate for generations to come. Economists have been warning against this potential for decades, and China's booming economy in the new millennium has very much made it a reality.

Yet, simultaneously, foundational shifts are occurring in Chinese policies and attitudes toward the environment. At the same time that endangered species are becoming speculative commodities within China, consumers and citizens are pushing for reforms. China's ivory ban in 2017 was backed by strong domestic support and has since implementation reduced willingness to buy ivory within the country by nearly half.[44] In the aftermath of the COVID-19 pandemic, the wildlife trade more broadly has become a target of reforms with broad-based national support.[45] Yet these are not carbon copy Western environmental reforms; rather, they are based on a very different view of the environment, nature, and sustainability than is often assumed in a Western context. China's wildlife ban, for example, contains specific provisions permitting trade for medicinal purposes. Beyond this, captive breeding and plans for the rewilding of tigers, as well as rosewood plantations and reforestation, continue apace. Differences between Chinese and Western approaches to the environment—including the significance of captive and wild populations—have only begun to be explored in this chapter. Chapter 6 will continue the discussion, with a focus on rosewood plantations and reforestation. These deeper cultural, philosophical, and geopolitical differences collide in the debate over endangered species protection, yet few even realize that they are there.

Beyond a purely logical dilemma, conflicting underlying assumptions of what the goal of endangered species protection is and of what sort of endangered populations are worth saving shape the debate on both sides, entrenching opposing positions further. Not seeking to advocate for one side or the other, this chapter highlighted the conflicting assumptions that come to bear on the debate. Divisions between licit and illicit, wild and captive, and so forth, as mentioned before, are not so much thin and blurry as distinct and doubled, drawn completely differently in different places. These fundamental divisions cannot be legislated away. Acknowledging these differences is not tantamount to promoting a cultural or moral relativism, wherein any and all actions are potentially legitimate. Rather, such acknowledgment is grounded in the practical interest of the survival of the species.

The survival of the tiger in China, rosewood across the tropics, or any endangered species with deep Chinese cultural roots is not achievable without the cooperation of the Chinese people. What these species mean to them, including their use, cultural symbolism, and exchange value, will not simply fade away in the face of Western values. To save the tiger in China, it may be necessary to compromise and expand a Western understanding of what such an effort really means. That is, it may be necessary to acknowledge the value of the tiger in both its domestic and wild valences, as well as the capriciousness of privileging one over the other. This is true for rosewood as well. Although Western NGOs operating in Madagascar privilege protecting the oldest, most charismatic rosewood trees in the forest, their energies might be better welcomed by local recipients if they were to focus on investing in future planting. Of course, these options of protecting and planting rosewood are not mutually exclusive, yet there remains a disproportionate focus on the former as opposed to the latter. As Chapter 6 will show, large-scale rosewood plantations established in southern China and other countries in Southeast Asia serve as an alternative model for reforesting the landscape and meeting future demand. These alternative approaches are likely to enjoy better reception from countries such as those in southern Africa that prefer to focus on captive breeding and trade legalization over strict protectionism.

China's captive breeding programs in tigers and other endangered species, as well as the country's large-scale forestry operations as discussed in Chapter 6, demonstrate an emerging environmental movement that contrasts with the preservationism of the West. Growing environmentalism in China thus provides an opportunity both to reflect on the conflicting underlying assumptions that shape approaches to endangered species conservation and to move toward unification rather than bifurcation. Given recent legislation, such as the 2017 ivory ban and the recent revisions to the Wildlife Protection Law, China is serious about cracking down on the illicit wildlife trade. Yet China's recent legislative history also demonstrates the country's focus on captive breeding, rewilding, and legalized trade as a critical way to supplement wild populations and meet future demand for certain cultural products. There is a clear desire to collaborate with expertise across the globe on this approach. African countries pushing for legalized trade of elephants and rhinoceroses offer a prime example. Rather than doubling down on strict protectionism, conservationists might take this opportunity to reflect upon Western perspectives on the illicit and the unnatural that are far from universal.

❋ 6 ❋

pluralizing environmentalism

In September 2019, the largest private-sector tree-planting initiative in China won the United Nations Environment Programme's Champions of the Earth award. The initiative, known as Ant Forest, is run on China's largest digital platform, Alipay (founded by Jack Ma, one of China's most famous billionaires). Ant Forest allows users to develop a type of individual carbon accounting based on their daily activities. Users monitor their carbon footprint and report low-carbon activities, such as recycling, riding a bike rather than driving, or making payments online rather than on paper. For each activity, they receive a certain number of green "energy points," which help grow a virtual tree in their profile and eventually a real tree in arid regions of northern China. Users can track the progress of their own virtual tree, as well as others' in their network, and when their real tree is finally planted, they can view its location via satellite. Initially launched in 2016, Ant Forest now has more than a half billion users and has planted more than one hundred million trees, reportedly reducing carbon emissions by nearly eight million tons.[1] The platform combines cutting-edge fintech innovation with nearly all aspects of contemporary everyday Chinese consumerism, while at the same time being grounded in one of the most ancient forms of environmentalism in China: planting trees.

Tree planting in China is perhaps the longest-standing environmental tradition, dating back centuries, if not millennia, and still very alive today. In the thirteenth century, already facing a largely deforested landscape, residents in southern China initiated large-scale reforestation projects alongside a sophisticated system of speculative trading in timber market futures established to finance the massive projects and hedge losses.[2] In this fashion, large-scale tree cutting and tree planting have gone hand in hand for centuries. Reforestation schemes continued on and off throughout the Ming and Qing Dynasties. Since the rise of the People's Republic of China in 1949, planting trees has become the most featured environmental activity of China's leaders. In the 1950s, Mao Zedong initiated large-scale afforestation programs (again, alongside massive tree cutting) to protect the North China Plain from advancing desertification.[3] Shortly after Mao's death, with the inauguration of Deng Xiaoping as the new paramount leader in 1978, construction began on the largest and most ambitious tree-planting campaign in history: the Great Green Wall, spanning all of northern China and including territories overlapping with Alipay's Ant Forest. Planned to be completed in 2050, the Great Green Wall continues today, now alongside myriad other tree-planting projects such as the nationwide Grain for Green program, China's first Greta Thunberg–inspired Fridays for Future (unsurprisingly, based on planting trees), and Alipay's Ant Forest. As a consequence of these initiatives, China plants more trees than the rest of the world combined and is the leading contributor to "global greening" trends.[4] Over the past two decades, China has undertaken the greening of an area of land one-quarter the size of the Amazon, mostly through increased forest coverage.[5] Chinese government agencies, along with the general population, see planting trees as one of the most promising ways to combat climate change, desertification, and pollution on a global scale.

All of this tree planting, however, is not without its controversies. Indeed, large-scale tree planting—including reforestation, afforestation, and ecosystem restoration—has become one of the hottest and most controversial topics in global climate and environmental debates over the past five years.[6] A 2019 article in the journal *Science* sparked much of the controversy by suggesting that planting trees across 0.9 billion hectares of the planet is one of the cheapest and most feasible means of combating climate change, especially in light of the massive political barriers to reducing fossil fuel consumption. The researchers in the study note that large-scale tree planting is not just one climate change solution but "overwhelmingly the top one," with "mind-blowing potential" to reduce emissions. According to their models, it would be "more powerful than all other climate change solutions proposed."[7]

The research has initiated a flurry of papers on the topic, some hotly contesting the findings, others demonstrating the beneficial climate mitigation impacts of tree planting already underway—mostly in China.[8] Aside from the question of the degree to which tree planting can deliver carbon mitigation benefits, however, is the deeper question of the social and ecological costs of such massive tree planting. While newly planted trees may be able to draw down atmospheric carbon stocks, are they also drawing down the water table, displacing communities, or threatening local biodiversity? These questions are most vociferously raised in relation to China. China's artificial forests are feats of large-scale ecological engineering that, many argue, do little to preserve prehuman ecosystems and deliver biodiversity benefits. They have been labeled "expensive band-aids," "green deserts," and failed attempts at "afforestation by monoculture."[9] Socially and ecologically, these massive projects are often viewed as downright disasters.

When it comes to rosewood, however, tree planting in China does protect biodiversity—just not through the typical Western approach. Chinese entrepreneurs, investors, and government agencies are beginning a new practice of planting rosewood trees to reforest the countryside and meet future demand, similar to the rosewood plantations started by Chinese Malagasy families in northeastern Madagascar, but at a much, much larger scale. As the price of rosewood furniture grows, it turns out, so too does the value of the trees in the ground. Thus, while China's booming rosewood market has triggered the devastation of forest resources across the globe, it has also triggered the proliferation of rosewood plantations as a related form of investment. Long-term investments in rosewood plantations have become increasingly common across southern China since the speculative dynamics of the wood became apparent around 2005. In China's southern provinces, where rosewood grew in large quantities long before the excesses of the Ming Dynasty, hundreds of thousands of hectares of rosewood are being planted by entrepreneurs and government bodies alike as a cultural, economic, and ecological investment.

These plantations are not simply endless rows of trees waiting to grow. Rather, following a long Chinese tradition of "dressing" the forest as though with clothing, rosewood plantation owners cobble together a diverse collection of flora and fauna that can generate income in the short term and exist harmoniously with the trees while they mature. These supplementary ventures are known as "understory economies" (*lín xià jīngjì*). Providing short-term returns to fund the long-term growth of the trees, understory economies rely on the price premiums of the cultural goods they provide: shade-grown traditional teas, organic free-range chickens, herbs for Chinese medicine, essential oils and incense for fragrance and health, artisanal honey and waxes, and tree saplings to be sold to other investors establishing their own plantations. As a whole, these forest plantation enterprises entail somewhat of a risk: they depend on the speculative value of rosewood. But mitigated by the short-term returns of the understory, and cushioned by the cultural and ecological appeal of rosewood and other niche commodities, the scales are tipping in their favor.

China's forestry projects—in terms of both large-scale tree planting to prevent natural disasters and rosewood plantations and their understory—are unparalleled around the world. Whether they are considered a success or failure, no other country is pursuing such initiatives in a comparable way or at a comparable scale. Moreover, as I argue in this chapter, these efforts are not only unique to China but also indicative more broadly of an approach to the environment that prioritizes large-scale ecological engineering and cultural cultivation over the preservation of natural ecosystems. Large-scale tree-planting projects such as the Great Green Wall or Alipay's Ant Forest are geared not toward preserving nature but rather toward tackling the extreme environmental threats of desertification, flooding, pollution, and climate change. Their aim, whether successfully achieved or not, is to quite literally *build* a sustainable future—that is, a future in which China will persevere and prosper despite existential threats such as climate change. Rosewood plantations, too, are more about propagating the species for future use than preserving it naturally in the wild.

Building a sustainable future through forestation (*zàolín,* literally "making" or "building" forests) is in many ways the opposite of preserving nature. This is not to say that one approach is better than the other, but that each retains a different allure for different people in ways we need to understand much better. This is also not to say that environmental movements in China such as forest-building are "pure" (i.e., motivated only by ecological considerations or some deeper ecological morality), especially when it comes to large-scale interventions. As emphasized throughout this book, the Chinese government is first and foremost driven by maintaining political control and social stability. Delivering environmental reform is not a motivation in itself but rather a response to demands from its citizens triggered by massive environmental degradation that has gone on for many decades. Responding to growing public demand for environmental improvement has become one of the government's primary ways of establishing legitimacy (and, consequently, maintaining political control and social stability).

Chinese government bodies, in other words, are interested in environmentalism not for some deeper philosophical or moral purpose but because the environment is undeniably a domain that must be tackled in order for China to be strong and prosperous. Rising environmentalism in China thus does not stem from government initiatives directly but from the demands of the people that have become to a certain extent, as discussed further in this chapter, codified by the government through the broad umbrella of what has come to be called "ecological civilization."[10] As an environmental ideal, ecological civilization is much more about building a future in which humans live in harmony with their wider surroundings than about protecting nature from human encroachment.

The difference is crucial. If our goal is the continued existence of endangered species in various parts of the world, better understanding of how people in those various parts of the world know and understand these species—as well as the environment more broadly—is imperative. This is even more the case when it comes to China because of the country's nonadherence to Western conventions and its increasing global influence. More than *protecting* or *preserving* natural ecosystems, Chinese environmental governance aims to "*construct* an ecological civilization" and "*build* a beautiful China." Compare these slogans, for example, with the (largely Western) charge to "save the planet" or the declaration that "nature needs half." They differ at every step: building and constructing versus saving and protecting, China and civilization versus nature and planet. As encapsulated in these slogans, China's environmental initiatives are based on a fundamentally different vision of a sustainable future—a vision that provides space for, even necessitates, large-scale human intervention.

China's approach to forestry reflects the country's unique environmental aspirations. Biodiversity conservation and nature preservation are not explicitly addressed through these projects—or at least not in a conventional preservationist sense—because these environmental goals are largely of Western origin and have not inundated Chinese culture as thoroughly as they have the West. Indeed, the concept and ideal of preserving a nature outside of humanity did not really exist in China until the early twentieth century, when it was introduced from the West.[11] Even now, the aspiration of preserving nature apart from humans lacks the widespread appeal it has gained in the United States and Europe (although the appeal is growing, especially in the younger generation). In the Chinese context—where it is difficult to even imagine what a prehuman natural ecology might have looked like—environmental aspirations are quite different.

This chapter examines China's approach to the environment—constructing and building more than preserving—with a focus on rosewood plantations and reforestation more generally. The chapter begins with China's general reforestation efforts, ranging from small local initiatives to reforest hillsides to large-scale ecological engineering projects that span the nation. These efforts have resulted in a vast, yet underacknowledged, arboreal transformation that rivals the country's more notorious urban transformation. The chapter then turns to rosewood plantations in particular, which have become increasingly common in the past decade as a result of endangered species speculation. As with large-scale ecological engineering, China's rosewood plantations are intended not to recreate a pristine forest but rather to cultivate and build a sustainable future. Using rosewood and reforestation as examples, the goal of the chapter is not to support or oppose China's rising environmental movement, nor to judge it as genuine or insincere, but simply to shed light on its unique cultural underpinnings, which are so often ignored. This, in turn, will allow for long overdue reflection on the unique cultural underpinnings of Western environmentalism that, as universal as it presumes to be, is not in fact shared by most people on this planet.

arboreal transformation

China's "great urban transformation" is well documented.[12] It provides the archetypal story of contemporary China: now more than half of the country's population lives in urban areas with no fewer than fifteen "megacities" (population of more than ten million) scattering the eastern seaboard. Alongside China's rapid urbanization, however, another transformation has been occurring outside the city limits (Figure 6.1). This transformation is much less recognized—both inside and outside of China. My family members in Shanghai who travel to rural provinces are surprised to find places they visited as youth now unrecognizable,

Figure 6.1. A slope in Hunan Province before 2001 (top) and in 2007 (bottom) after the implementation of the Grain for Green program, one of China's largest reforestation / afforestation programs. Jien Zhang, Tianming Wang, and Jianping Ge, "Assessing Vegetation Cover Dynamics Induced by Policy-Driven Ecological Restoration and Implication to Soil Erosion in Southern China," *PLoS ONE* 10, no. 6 (2015): e0131352.

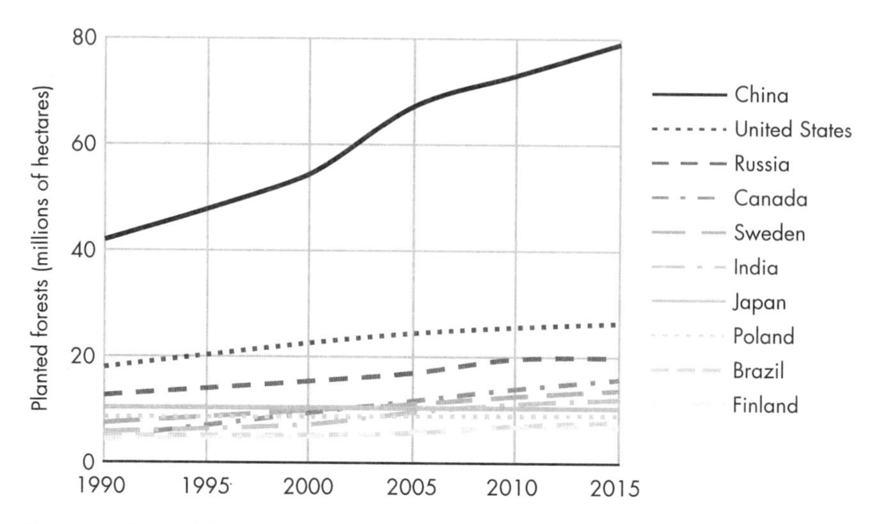

Figure 6.2. Planted forest coverage (millions of hectares) for the top ten tree-planting countries, 1990–2015. *Data source:* FAOSTAT 2019.

covered in trees. China's tree-planting campaigns have left the country with far more planted forests than any other country in the world (Figure 6.2). China's Great Green Wall and Grain for Green program provide the two most iconic examples, together involving replanting efforts that span from the far north to the far south of the country. In addition to these mega-forestry projects, nearly every new development project is accompanied by large-scale tree planting, with typically at least one-third of the footprint consisting of green space. China might be best known for the major megalopolises dotting its eastern seaboard, but massive reforestation is happening at an arguably larger scale through some of the largest state-led and private environmental projects in history.

In modern China, large-scale tree planting began with Mao Zedong. Nearly all of Mao's state-making activities involved waging a "war against nature"—aptly summarized in the title of Judith Shapiro's famous book on the topic.[13] Yet, as forceful as his war against nature may have been, Mao often found himself on the defensive when it came to the advancing Gobi desert and the massive sandstorms it spawned. Fending off catastrophe, Mao ordered tree planting over one hundred million hectares of degraded land as "only a beginning."[14] By 1958, he called for one-third of the entire cultivated area of China to be planted with trees.[15] The largest among Mao's projects was what would later become known as the Three-North Shelterbelt Project (referring to the three northern regions of the north, northeast, and northwest)—more colloquially known as China's Great Green Wall. Planned to conclude in 2050, this massive ecological engineering effort will span more than seventy years and cover approximately one-third of the countryside in northern China, providing one of the most extreme examples of long-term state planning—even by Chinese standards.[16]

Mao's successor, Deng Xiaoping, was also a notorious proponent of tree planting.[17] He promulgated legislation requiring that every Chinese citizen with the ability to work plant three to five trees per year. As with Mao, Deng's goal was not to restore a natural ecology but to provide materials for production and prevent environmental catastrophes associated with rampant deforestation (flooding, sandstorms, and so forth). Deng's Capital Greening Committee, in charge of enacting the legislation, further mandated that "if you plant trees, you must choose those that grow fast and can be made into materials."[18] Alluding to the various layers of forest production, Deng noted that the forest should be "dressed" like a person: "not only must it have clothes, but also shoes and accessories."[19] Here we see an early version of the understory economies that now supplement China's rosewood plantations on a large scale.

Most recently, President Xi Jinping has reemphasized the importance of tree planting, along with environmental protection more generally. China's 13th Five-Year Plan (2016–2020) reports that the country's forest coverage goal of more than 21 percent by 2015 has been met, and the forest stock goal has been exceeded by nearly one billion cubic meters. To reach these goals, Xi reassigned sixty thousand military troops from northern border security to "build" three new state forests, planting 6.66 million hectares—an area the size of Ireland—in 2018 alone.[20] The latest Five-Year Plan (2021–2026) calls for increased forestland coverage and restoration as a means to reach China's climate goals of peak emissions by 2030 and carbon neutrality by 2060.[21] Specifically, China plans to plant over 33 million hectares of forests and grasslands during this five-year period (an area of land just under the size of Ireland every year).[22] Planting trees, in China and increasingly abroad, will play a substantial role in reaching the country's emissions targets.

Such massive tree-planting efforts—from Mao to Xi—have received understandable criticism. Regarding the Great Green Wall, survival rates have been pitiful, especially in the project's early years, and the environmental impacts remain deeply contested.[23] The project is seen as either the largest environmental project in history or one of the largest human-made environmental failures in history, depending on whom you ask.[24] It has been estimated that less than 30 percent of China's planted trees actually survive to this day.[25] This is in stark contrast to the Mao-era tree-planting campaigns, for which official statistics from the Communist Party claimed survival rates of more than 70 percent.[26] Overall, China's official statistics indicate consistent growth in forest cover since

1950, but scholars from the United States and Europe have cast doubt on these numbers.[27]

A second critique leveled at China's forestry efforts—more subtle in nature—questions not necessarily their longevity but their ecological suitability. Do these newly planted trees contribute to the "natural" ecology? In other words, the trees may reduce erosion and enhance water quality—they may, in short, *survive*—but do they support the biodiversity of the natural landscape that preceded them? The question of supporting a predisturbance ecosystem arises frequently when considering China's forestry. Can we call such rampant tree planting of fast-growing species "environmental"? Or is it a type of large-scale ecological engineering that has little regard for predisturbance ecosystems and biodiversity?

China's large-scale experiments with tree planting clearly diverge from visions of pristine nature and a predisturbance ecology. In many ways, such large-scale tree planting is no different from the country's large-scale urbanization: impressive feats of engineering the landscape to Chinese standards. In this view, it is hard to call such tree-planting efforts environmental. What I will argue in the pages that follow is that China's tree planting is indeed environmental, but it is based on a particular approach to the environment that does not value preserving nature apart from humans. Sustainability and longevity are the goals, and improvements have been made in this regard: using more bushes and shrubs than trees as the water table requires, not planting monocultures so as to avoid susceptibility to disease, choosing species that have economic benefits as well as environmental, even cordoning off large areas to regrow by themselves through a system of "ecological red lines." But recreating a natural landscape is not a part of the plan. Before examining this argument, however, let us first turn to China's rosewood plantations—a different type of large-scale tree planting that intertwines China's cultural, economic, and ecological aspirations even more deeply.

rosewood and the understory

Few of China's artificial forests have the cultural allure of a rosewood plantation. Given the speculative value of the wood and the assortment of companion species—teas, chickens, honey, and herbs—that compose its understory, China's rosewood plantations are unique. State and private investors are planting rosewood for a confluence of economic, ecological, and cultural reasons that are unique to this group of trees. Despite their growing appeal, however, rosewood plantations still lag far behind eucalyptus and other fast-growing species in terms of surface area coverage.[28] Yet, in part as a result of new provincial policies that discourage fast-growing tree species, rosewood plantations are on the rise.

Following our trip to Zhongshan and the rosewood mega-mall discussed in Chapter 1, my husband and I traveled with researchers from the Chinese Academy of Sciences to visit three such rosewood plantations in southern China, just outside Guangzhou city limits. Escaping the urban sprawl was not easy, especially in the snarl of midmorning traffic, but as the high-rises and freeways gave way to rural townships and country roads, the landscape in this part of southern China surprised us with its lush vegetation and verdant hills. Soon enough, however, these "forests" took on a subtly unsettling quality. The height and size of the trees and the spaces between them became highly regular, with an effect similar to traveling inside of a low-budget, procedurally generated computer simulation. The researchers accompanying us explained that we were surrounded by eucalyptus plantations and that the one we were driving through at that moment was a mature grove—all of the trees were five or six years in age—ready for harvesting and processing into cardboard. The scale of these agroforestry operations slowly became apparent as we continued our course, winding through what seemed to be endless repetition of a single landscape.

Eucalyptus, as our host researchers explained, typically grow for five to eight years and then are cut and processed into pulp for cardboard (Figure 6.3). As cardboard boxes, the wood pulp will eventually make its way back to Guangzhou and other urban centers across China. With such short rotations, eucalyptus often results in more negative than positive environmental consequences. It contributes to soil degradation and poor water quality while providing marginal returns. Eucalyptus plantations earn considerably lower returns than agroforestry enterprises based on more lucrative species. Yet they still represent a large economy in southern China, with a combined value of US$430 million in 2015.

In an effort to dissuade landholders from planting eucalyptus, the governments of Guangdong and Guangxi Provinces have recently established incentives for planting longer-rotation species. Growers who buy permits for cutting their trees (all tree cutting in the province requires a permit) now get money back if, afterward, they plant noneucalyptus species as a replacement. This policy represents a clear change in comparison with China's earlier approach of promoting reforestation of the species.[29] Considered some sort of miracle tree with rapid growth and substantial environmental benefits, eucalyptus was promoted through afforestation policies well into the 1990s.[30] After 2000, however, it gradually became clear that eucalyptus forests were not as environmentally beneficial as once thought.

Figure 6.3. Pulp from eucalyptus trees to be made into cardboard in Guangdong Province, March 2018.

Given that eucalyptus leaches nutrients from the soil and increases erosion, the government was looking for other tree species to take its place.

This is what my husband and I learned as we pulled up to our first rosewood plantation of the day—a thousand acres of intercropped rosewood and sandalwood on rolling hills, established by a textile-manufacturing mogul who, for the sake of anonymity, we will call Mr. Bai. We arrived at Mr. Bai's plantation as the repetition of eucalyptus in the landscape finally gave way. Veering off the highway onto a dirt road, we drove down to a clearing by a small lake. Here again were the regularly spaced plantings of young trees but with crucial differences. Approaching on foot, led by the researchers, we walked amid alternating plantings of blonde-barked rosewood (*Hǎinán huánghuālí,* with the scientific name of *Dalbergia odorifera,* native to China's Hainan Island) and darker sandalwood (*Santalum album,* native to Australia and India). We learned that we were standing at the edge of one thousand acres of rolling hills with intercropped rosewood and sandalwood. At ten years of age, both tree species were around the same height, but the rosewood is expected to grow quite tall, and the sandalwood does not mind growing in its shade.

Despite being in the middle of an agroforestry enterprise, the plantation felt airy and rather gardenlike, with dirt trails cutting through the groves and along the lake. A concrete structure in the middle housed offices and a small exhibition space, showing off rosewood and sandalwood products. Walking down one of the paths, we could hear the crow of a rooster. Then, through the trees, we spotted hens picking their way around the dry foliage. A moment later, we noticed what appeared to be intravenous fluid bags attached to some of the trees, which fed down to what looked like small nests fashioned out of black plastic netting. "This is the understory economy," the lead researcher explained.

The "understory economy," as discussed earlier, refers to what is produced and sold in the microhabitat under the shade of the precious hardwoods, which take much longer to mature. For example, most rosewood cannot be harvested for at least twenty years and is not truly desirable for furniture until it is at least thirty-five to fifty years old, so other agricultural enterprises supplement the plantation, providing income and value while the hardwoods mature. After a few decades of good growth, *Hǎinán huánghuālí* can be worth US$1,000 per kilogram—double the price of silver. But the plantation owner must financially support its growth up until that point. A hundred-year-old tree of the species (which no longer exists, either in the wild or under cultivation) would theoretically be worth its weight in gold.[31] With such extravagant prices, a young rosewood forest is worth a future fortune. But the understory is what makes waiting for that future fortune possible.

The sandalwood trees that flank both sides of the rosewood provide an obvious example. Like rosewood, sandalwood is a rare, precious hardwood that has earned record returns in the past decade. The sandalwood essential oil industry, like the rosewood furniture industry, has been on the rise because of Chinese demand. The wood can be made into essential oils that sell for about US$3,000 per kilogram—about five times the price of silver.[32] For their oils alone, each tree could earn at least US$1,500. In the meantime, the smaller pruned branches can be sold to make incense, and the dried buds can be sold to make tea.

Initially, Mr. Bai had attempted to grow only sandalwood, planting most of his thousand acres with only this tree. Yet after two or three years of promising growth, all of his trees mysteriously died. Working with researchers from the Chinese Academy of Sciences to figure out why, they eventually discovered that sandalwood planted by itself soon depleted soil nutrients, but with a particular companion crop, the tree could grow quicker and for a longer period of time. This companion crop turned out to be *Hǎinán huánghuālí,* one of China's most famous rosewood species and far more lucrative at maturity than sandalwood. Planted together, the two species thrive, with sandalwood providing valuable short-term returns in the meantime.

The epiphytic recipients of the intravenous drip we witnessed provide another example. These epiphytes—orchids, in fact—are a famous Chinese herb (*Dendrobium nobile*) valued primarily for its anti-inflammatory properties and used in traditional Chinese medicine for centuries. The orchid uses the bark of the hardwoods as a substrate, and the drip system provides a steroid to fortify the orchid's growth. When harvested, the orchids are worth US$16 per kilogram (fresh). Compared with rosewood, the orchid is a relatively fast growing commodity that can be cultivated and harvested seasonally. In addition to this orchid, chickens provide a valuable addition to the understory economy. Preying on insect pests as they stalk the property, the chickens at this plantation were fine examples of local breeds. Also, because they range freely, their meat and eggs are more valuable than industrially produced supermarket products.[33]

At the end of our tour, we finally met Mr. Bai. Although one might consider him a millionaire industrialist, in demeanor and dress he was indistinguishable from the groundskeepers around him. Having amassed his fortune from humble origins, Mr. Bai was keen to return part of his earnings to the rural countryside in which he grew up. He was also quite eager to talk about his investments in the plantation and his general interest in rosewood. How and why did these diverse living things come to be cultivated together in one place? In 2005, he signed a thirty-three-year lease on the land from a local collective. The land was previously planted with fruit trees and not intensively worked. His first thought was to plant eucalyptus, but learning of the environmental consequences of the tree and seeking to differentiate his investment from the pack, he decided upon other slower-growing, higher-return species. With the wealth he had amassed over the previous two decades, he now wanted to secure his legacy. So, instead of planting a six-year eucalyptus rotation for cardboard pulp, he invested in slow-growing precious hardwoods that can require upward of fifty years to mature.[34]

Now, with roughly ten-year-old trees, his plantation was doing quite well. Three years prior, Mr. Bai told us, a company from Hong Kong had offered to buy the plantation for more than US$23 million—more than five times the price of his initial investment. Mr. Bai had declined, asserting that his plantation was in fact worth well more than US$30 million, but he was not, after all, interested in selling. Having earned a great deal from his textile factory, he was already a wealthy man. Besides, with a thirty-three-year lease costing about US$16 per acre, the medicinal orchid he was growing could alone cover his rent. As an early investor, he benefited from much cheaper rents than investors now confront. On generous terms by today's standards, the same lease would now be worth considerably more and would be much more difficult to obtain.

Later in the day, we visited a second plantation owned by a man we shall call Mr. Xu. Mr. Xu was a retired government forestry official in his seventies who carried himself with scholarly gravitas. He was likely even more wealthy than Mr. Bai, although you would never guess it from his work attire and modest demeanor. At his plantation, we saw tea plants growing in the shade of a grove of nine-year-old rosewood trees, again *Hǎinán huánghuālí*, bordered by sandalwood and another rare species of rosewood, *Dalbergia cochinchinensis* (Siamese rosewood). The tea, like the other understory economies, added value to the land in more than one way. At only 330 acres, Mr. Xu's plantation earned more than US$300,000 per year from the tea crop alone, including the sale of tea saplings, leaves, and mature trees. In 2015, Mr. Xu was offered more than US$3 million for the purchase of his plantation. He declined the offer, insisting that his rosewood trees alone were worth closer to US$5 million at the time and likely much more in the future. This plantation, like Mr. Bai's, also benefited from a surprisingly low lease price.[35]

Certainly, tea sales make a capital contribution to the plantation economy, but growing the tea immediately impacts the environment of the plantation as well. The fragrance of the flowers and the spacing required to grow the tea together with the trees gives these plantations the feel of a country garden. They are undeniably pleasant spaces for an afternoon stroll, which their owners no doubt take advantage of as one of the perks of their business. Additionally and perhaps more significantly, the tea has real cultural value. Like rosewood, tea is more than just a commodity in China; it is intimately connected to a set of social rituals and habits that hold profound cultural and historical meaning. To be a cultivator of tea is perceived as a worthy and gentlemanly pursuit, especially in juxtaposition with the eucalyptus-farming neighbors, growing a foreign stock of trees bound for foreign lands. Indeed, most of the understory economy commodities we observed had far greater cultural value than simple use value. In our conversations with the plantation owners that day, it became crystal clear that they found far deeper satisfaction—and far greater value, although not in a purely economic sense—in their rosewood plantations than they would have gained from growing eucalyptus for cardboard pulp or any other purely economic pursuit.

When we asked Mr. Xu at what point he planned to harvest his rosewood, he looked a bit surprised. It was evidently not something he was particularly eager to talk about. After some hesitation, he replied that he supposed he would have to harvest some trees when his initial lease expired, in order to pay for the inevitable rent escalation. Like Mr. Bai, Mr. Xu had contacted the Chinese Academy of Sciences to assist with his investment. Alluding to the irony, the researchers from the academy joked that, as a retired forester for the Communist Party, Mr. Xu in fact knew much more than they did about cultivating rosewood. He was, it became clear, much more interested in having his trees grow than cutting them down. But certainly, when the time came, cut them down he would—at least the majority of them.

Our final visit of the day was to a government-sponsored demonstration plantation, run by the researchers who escorted us. In order to encourage the types of forestry investments made by Mr. Bai and Mr. Xu, the Chinese Academy of Sciences' Research Institute of Tropical Forestry established a number of demonstration plantations throughout southern China. We visited the largest among them: the Beiling Mountain rosewood plantation, containing more than two thousand hectares of rosewoods, sandalwoods, and Chinese fir (see Figure i.3). The plantation is intended to serve as a model for landholders in the region. It stands adjacent to the much older Dinghu Shan Nature Reserve—China's first nature reserve, established in 1956. The Dinghu Shan Nature Reserve was created as a monument to the region's unique ecology, a meeting point of deciduous evergreen and broadleaf trees. Right next door, the Beiling Mountain rosewood plantation is intended to demonstrate something else. In contrast to the nature reserve, the rosewood plantation demonstrates what humans can do within these broad ecological constraints. Using the region's unique ecology, this massive government-sponsored rosewood plantation demonstrates how to live off the land in an era of booming markets in niche luxury goods, where certain agricultural products have become a type of hypercommodity through their speculative future returns. It provides a unique demonstration of what many Chinese people have come to call "ecological civilization."

"constructing" an ecological civilization

China's rosewood plantations, their understory economies, and the vast tree-planting campaigns that cover the countryside more broadly are all empirical manifestations of the growing discourse on "ecological culture" (*shēngtài wénhuà*)—and the broader conception of an "ecological civilization" (*shēngtài wénmíng*)—toward which China currently aspires. Ecological culture and ecological civilization are concepts that have emerged forcefully in China's environmental discourse over the past decade. Ecological civilization has gained particular salience, now replacing many references to what was formerly called "sustainable development" (*kěchíxù fāzhǎn*). When analyzed together, the concepts of ecological culture and ecological civilization help distinguish what is different about China's approach to the environment from the West's.

Although closely related and sometimes used interchangeably, ecological culture and ecological civilization retain minor differences. The genealogy of the term "ecological culture" can be traced to the former Soviet Union as an emerging Marxist-Leninist concept that posed ecology as part of a mature socialist future. The term quickly caught on in China, where it would end up becoming much more influential. In 1987, in his opening speech at the National Conference on Eco-agriculture, the famous ecologist Ye Qianji called for the need to "pioneer the construction of an ecological culture."[36] For Ye, "ecological culture" referred to a mutually supportive relationship between humans and their wider ecology, or, in his words, "a harmonious relationship where humans and nature are mutually supportive." He concluded that "from the perspective of an ecological culture, a truly civilized era has only begun."[37] More recently, the influential Chinese economist Mai Yining described ecological culture as a type of development "built on a positive foundation of the cycles of the ecosystem"—in other words, a holistic development that views civilization as deeply engrained within ecology.[38]

"Ecological civilization" has come to surpass "ecological culture" in its usage, now serving as the broad rhetorical and policy umbrella under which nearly all environmental initiatives in China are situated. With similar socialist origins, ecological civilization was popularized by then president Hu Jintao in 2007. In 2012, the "construction of an ecological civilization" (*shēngtài wénmíng jiànshè*) went on to become one of the Chinese Communist Party's paramount objectives, and it was written in as a constitutional principle in 2018.[39] Both terms—"ecological culture" and "ecological civilization"—now hold "particular sway in China's environmental discourse" and are used by politicians, scholars, and planners to define their visions of an appropriate future.[40] They are in many ways meant to ground China's emerging environmental movement in distinctly Chinese terms. They are also beginning to gain global salience, shaping global environmental debates as well.[41] According to a recently translated volume on Chinese conceptions of the environment, these concepts "will undoubtedly be the theme of the twenty-first century."[42]

At first glance, "ecological civilization" may not strike one as very different from "sustainable development," the term more frequently used in a Western context. Ecological civilization has been referred to as "sustainable development with Chinese characteristics" or "a Chinese concept of sustainable development."[43] One major difference, however, is that ecological civilization goes beyond the rather technocratic approach of sustainable development, bringing politics and culture more centrally into discussions of the environment. Ecological civilization disrupts the conventional triangulation of *social-environmental-economic* factors characteristic of sustainable development, adding often overlooked political and cultural dimensions.[44] Politically, the role of the state in responding to environmental crisis and building an environmental future is much more prominent in discussions of ecological civilization than of sustainable development. Central authority and planning are integral features of ecological civilization construction. Culturally, the concept of ecological civilization draws on ancient Chinese philosophies that emphasize balance and harmony, ranging from Confucianism to Taoism. By including political and cultural elements, the concept of ecological civilization is grounded in distinctly Chinese terms.[45]

In comparing the political and cultural underpinnings of ecological civilization in China with environmentalism in the West—specifically in the United States—a number of differences become apparent. To begin, environmentalism in the United States does not have the same political consensus that ecological civilization enjoys in China. Ecological civilization represents *the* contemporary approach to the environment in China. Although there are various interpretations of what specifically an ecological civilization entails and how to achieve it, there is nonetheless general agreement that it is the overall goal of Chinese environmentalism in the twenty-first century. This is fundamentally different from American environmental movements, which can be considered either a marginal or primary political agenda, depending on the administration. Environmentalism in the United States plays a less certain role in central politics. But of course this is true for *all* governance domains when comparing Chinese and American politics, environmental or otherwise. This difference, in other words, is not characteristic of Chinese or American environmentalism but rather of their political systems writ large. On top of this, many might rightfully argue that China's supposed "political consensus" is highly imposed and obscures a great deal of dissent simmering within.

Beyond politics, however, are the different cultural underpinnings and ideological outlooks of ecological civilization and American environmentalism. Although these differences may be more subtle, they turn out to be quite consequential. Let us begin with the cultural and ideological underpinnings of American environmentalism. As varied and heterogeneous as this category is, there are two opposing strains of American environmentalism that can be identified on the basis of two uniquely Western philosophical traditions.[46] Both strains are anthropocentric in that they are based on inescapably *human* cosmologies, but whereas one of these strains embraces its anthropocentrism, the other attempts to cast it aside.

The first strain is rooted in the *utilitarian* mentality of Gifford Pinchot, which emphasizes the conservation of resources for human use. It builds on classic Enlightenment thought, acknowledging the benefits of rationality and "progress" associated with transforming nonhuman natures, but it cautions that humans must be modest and conservative when repurposing nature to meet their own ends. Sustainable development is often criticized for falling into this category. As a guiding environmental principle, sustainable development emphasizes human use of natural resources equitably over space and time, yet it ignores the inherent value of nonhuman nature for its own sake. This line of thinking inspires certain brands of "techno-optimism"—the idea that human progress can solve our global environmental woes—an ideology whose followers historian Charles C. Mann labels "Wizards" (as opposed to "Prophets").[47]

This is where the second strain of environmentalism enters, the more "Prophet-like" strain, often defined in opposition to the first. This strain highlights the follies of human intervention and advocates nonintervention as the appropriate response. The result is a *preservationist* mentality, that of Ralph Waldo Emerson, Henry David Thoreau, and John Muir, which emphasizes the *transcendental* value of nature beyond or above human utility. It purports to be an eco-centric (as opposed to anthropocentric) environmentalism and can be traced to a unique reversal in Enlightenment thought. By the late eighteenth century, Romanticism and, later, transcendentalism began attributing value to nonhuman nature—nature not for human use but for its own sake. That which remained untouched by humans in fact became *more* valued and esteemed than the human world.

Untouched wilderness, once desolate and downright terrifying, became cherished, "the best antidote to our human selves."[48] And thus emerged a peculiarly Western approach to valuing pristine nature, untouched by humans.

Both Enlightenment utilitarianism and the Romantic and transcendental reactions to it (and, consequently, both anthropocentric and eco-centric environmentalism) are structured on a foundational dichotomy between human and nonhuman, nature and culture.[49] This basic opposition—so foundational to Western thought—remains today, providing the conceptual scalpel with which the first and deepest incision has been made and giving form to a distinctively Western model of the phenomenal world. The division is, as Jacques Derrida notes, "congenital to philosophy" and can be witnessed as early as the Greek Sophists (fourth and fifth centuries BCE, based on the *physis* / *nomos, physis* / *technē* distinction).[50] This generative opposition makes possible all scientific inquiry today as divided by the "natural" and the "social" sciences, as well as environmental ideologies insistent on preserving the nonhuman world. The Romantics and transcendentalists who have come to inspire contemporary eco-centrism did not question the nature / culture divide; they simply reversed its normative hierarchy. Nature for its own sake was considered more important than the culture of rationality and "progress" meant to exploit it.[51] Echoes of this dichotomy are clearly present in preservationist movements of the contemporary era, such as the "nature needs half" movement and the "fortress" conservation discussed in Chapter 4.[52] These are quintessential examples of the imposing "ideological position" of "protectionist NGOs" that the Southern African Development Community vehemently condemned (see Chapter 5).

Environmentalism in China is not structured by the same nature / culture opposition. Although there is certainly debate over what constitutes a healthy environment, this debate is not premised on the presence or absence of humans. As with most non-Western cultures, the distinction is not foundational to Chinese thinking. This has been the case since the establishment of the earliest Chinese parks and reserves, such as those that date back to 300 BCE. "These were not wilderness," historian Johnson Hughes writes in observation of their marked difference with Western reserves, but rather "gardens [in which] every bit is designed, and art exhausts itself to be indistinguishable from nature."[53] At that time in China, in fact, there was no word for "nature" as it is used in the classical Western sense of nature versus culture.[54] Rather, "nature" (*běnxìng*) referred simply to the inherent quality of a thing (its essence or most basic substance)—human nature, bear nature, tiger nature, the nature of a mountain, house, or building. Only during the twentieth century, after significant influence from Western powers operating within the country, did "nature" in the Western sense, meaning the world separate from humanity (*zìrán,* or sometimes *dà zìrán*), become a part of Chinese vernacular.

Although Chinese culture has to a certain extent adopted a Western concept of nature, and the word for "nature" in this sense is now widely used throughout the country with many not aware of its Western origins, the nature / culture opposition on which this concept is based has not so deeply penetrated Chinese ideas about the environment. The structuring oppositions of nature versus culture and nonhuman versus human do not determine everyday Chinese thought as they do in the West. Instead, the philosophies that guide Chinese ideas of the natural world and cosmic order do not readily differentiate human and nonhuman elements. A clear example is the Chinese saying *tiān rén hé yī*—frequently translated as "the harmony between humans and nature." However, the word translated as nature (*tiān*) in fact means "heavens" or "cosmos," again because the concept of nature as opposed to the human is a more recent addition to Chinese vernacular.

Another example is the concept of *qì*—the flow of energy that animates all things. This concept, which is foundational to Chinese thinking when it comes to the environment, does not follow boundaries of nature or culture. *Qì* flows through natural landscapes and built environments alike—forests, mountains, cities, skyscrapers. "This is not a nature apart from human culture," anthropologist Robert Weller writes, "but the flow that underlies both nature and human life."[55] Similarly, *fēngshuǐ,* or feng shui, the art of optimizing such energy flows, is not limited to the human domain, as commonly understood in the Western context of interior decorating. Feng shui specialists analyze the flow of *qì* in one's home, across a city block, or among trees in a forest through the same basic principles. The focus of these philosophical concepts is always balance and harmony, whether in nature or culture, and there is no meaningful delineation between the two.

Interestingly, Western astrology (rooted in pre-Enlightenment beliefs emphasizing correspondence and similitude), too, does not strictly follow a nature / culture opposition: the movement of the stars corresponds to one's fate, a process through which natural and human laws are indistinguishable.[56] But in contemporary Western society, astrology is an intriguing irrationality, an amusing if not believable pastime—that is, the exception that proves the rule. In China, this is not the case. Foundational concepts that transgress the nature / culture divide, such as *qì* and *fēngshuǐ,* shape everyday life and thinking about the environment. A house with good *fēngshuǐ*—"wind" and "water," as the word literally translates—is open to the forces of the universe, bringing good fortune. Eating fresh seasonal foods, too, will align one's internal system with cosmological rhythms (i.e., the rhythms of "nature") not because this food is "local" or "socially just" but because human bodies are believed to be subject to the same cosmological forces as the world around them. The cosmic forces that animate the tiger, the pangolin, and countless other plants and animals, granting vitality and strength, are believed by many Chinese people to do the same for humans if consumed. It is in fact this perceived cosmological unity—*tiān rén hé yī,* not structured by a nature / culture opposition—that lies at the core of Chinese wildlife consumption.

The absence of a nature / culture opposition also pertains to the concept of ecological civilization, as has been clear since the concept's academic debut with the speech of Ye Qianji at the National Conference on Eco-Agriculture in 1987. Subsequent scholars who have further developed and popularized the term, such as the famous Marxian economist Liu Sihua, have continued to emphasize its nonbinary approach. Through the concept of ecological civilization, Liu observes, "the ideas of *humanizing nature* and *naturalizing humanity* should be combined into an organically dialectical process."[57] As one of the foremost proponents of the concepts of ecological civilization, Liu speaks of building a Chinese ecological civilization based on an ecological culture yet also maintains that "this new idea of civilization's properties is not anthropocentric."[58] At the same time, one must admit it is not particularly *eco*-centric, at least not in the Western usage of the term.

Herein lies my argument: casting China's approach to the environment in terms of anthropocentrism or its constitutive opposite simply misses the point. Although Westerners may look upon Chinese environmental projects and declare they are indeed anthropocentric—massive tree-planting campaigns to protect (or control) humans in desert-threatened regions at the expense of nature, rosewood plantation owners looking only for profit rather than environmental protection—the reality is that anthropocentrism makes sense only in the presence of its obverse. It works as a charge only when contrasted with the logic and possibility of eco-centrism. Both anthropocentrism and eco-centrism as its obverse are possible only in a thought system that assiduously separates the human and the natural. Yet from a perspective that does not abide by such separation, neither preserving nature outside of the human nor developing resources only within the context of some predefined human use makes much sense. Quoting Xu Chun, another major proponent of ecological civilization, Liu reminds us, "neither social development that neglects nature nor a natural environment devoid of people should be the deeper meaning of sustainable development. Since sustainable development is development for people as well as nature, it is the development of man in nature and nature in man."[59]

In this light, critiques of China's planted rosewood forests and ecologically engineered forests take on new meaning. Yes, from a perspective that values biodiversity prior to human disturbance, China's artificial forests and plantations do not really count as forests. The natural ecology is not being restored but rather engineered anew. From this perspective, it comes as no surprise that China's artificial forests and plantations face sharp criticism for being "empty forests" that do little to contribute to biodiversity or revitalize the natural ecosystem.[60] From the perspective of ecological civilization, however, China's planted forests and large-scale ecological engineering projects look quite different. Harmonizing cultural traditions and natural forces, China's rosewood plantations provide an ecology that is in many ways preferable to whatever natural forest might have existed long ago. At their best, they demonstrate the harmonious interdependence of natural growth and the cultivator's intention—a step forward toward an ecological civilization. Similarly, large-scale engineered landscapes reproduce this logic at spatiotemporal scales more closely matching those of a *civilization*. The Great Green Wall, for example, operating on the ecosystem level and spanning two centuries (1978–2050), illustrates the sheer magnitude the concept of ecological civilization is intended to convey. As with rosewood plantations, these massive feats of ecological engineering are about "building" and "constructing" to achieve desired ecologies.

My point is not to defend these projects, neither rosewood plantations nor large-scale tree planting, as "truly" environmental after all, but to show that if they are at all environmental, it is not primarily in the Western sense of the word and thus should be understood to the extent possible outside of a Western framing. Moreover, precisely because ecological civilization is *not* structured by the uniquely Western nature / culture opposition, the movement may end up reaching far beyond China, potentially gaining an appeal abroad that strict preservationism lacks.

moving beyond "good" or "bad" China

When it comes to understanding China's role with respect to the global environment, debates focus on China as either a long-standing environmental threat or—more recently—an emerging environmental leader. On one hand, the international community criticizes China for overconsumption and environmental disregard. On the other hand, many within the international community increasingly look to China to fill a void left in the global environmental arena as Western countries draw inward. Sitting at the forefront of pollution reduction and renewable technologies, the largest proposed national carbon-trading regime, and now the largest economy within the global climate agreement, China has paradoxically become both the greatest global environmental threat and an emerging environmental leader.

This is true for forestry as well. China is the world's largest deforester but also, as we have seen, the world's largest tree planter. China contributes more than any other country to global greening trends. Increasingly, China is also exporting its model of large-scale forestry abroad. President Xi has asserted that the country will "promote afforestation via multilateral cooperation mechanism, such as the Belt and Road Initiative, so as to cope with global challenges, such as climate change, and to contribute its due share to global ecological security."[61] In addition to the Great Green Wall project in northern China, for example, the country has also agreed to assist with Africa's Great Green Wall, another massive reforestation project spanning more than eleven countries at the southern border of the Sahara Desert. Moreover, the Chinese government is assisting neighboring Southeast Asian countries in establishing rosewood plantations similar to those established domestically. When it comes to the Belt and Road Initiative, China is developing "green economic belts" based on tree planting. Starting in northwest China and extending into Central and West Asia, the belts will be created using poplar trees—a controversial species that some critique as fast-growing monocultures and others laud as having brought ecological and economic revival to arid regions in northwestern China.[62]

As many condemn China's global environmental footprint, indiscriminate logging, and insatiable appetite for rosewood and other endangered species, it is useful to look at the story from another angle. From this new perspective, we can see that China's approach to the environment is not particularly "good" or "bad" but fundamentally different from a Western approach, with profound consequences for global governance. This is not to essentialize Chinese or Western environmentalism but to emphasize that as heterogeneous as each may be, the two rely on very different cultural underpinnings. Whereas China's rosewood plantations and large-scale artificial forests are far from "pristine nature," they nonetheless embody a particular approach to the environment that is becoming increasingly influential in China and now globally. It remains to be seen whether this approach will be successful in any sense of the term. But, as this chapter has shown, not all countries use the same rubric when it comes to defining environmental success.

Appreciating the cultural specificities of China's approach to the environment allows for not only a more clear understanding of the country's environmental movement but also greater reflection on the cultural specificities of Western environmentalism. Certainly there are overlaps between Chinese and Western approaches to the environment and great variations within each, but there are also key differences that will be critical to global environmental governance moving forward. Far from universal, Western environmentalism has unique features—particularly when it comes to ideas of nature—that are not shared by most people on the planet.[63] Provincializing Western approaches to the environment allows for the consideration of other situated approaches, Chinese and otherwise. It allows for the *pluralization* of environmental approaches at the global level. This, in turn, will allow environmental debates to move past China as a "good" or "bad" environmental steward and onto a new horizon of reconciling Chinese and Western approaches to the environment. Without such reconciliation, there is little hope of collectively facing the looming environmental catastrophes of the twenty-first century.

conclusion

global Chinas of past and present

At some point in the 1990s, as rosewood stocks began to diminish throughout Asia, an enterprising Chinese trader stumbled upon an imperial timber reserve that had been hidden for centuries. Dozens of logs of the famous *zǐtán*—one of the most prized rosewoods of the Ming Dynasty and commercially extinct in the world today—were brought into circulation. At least, so goes the story circulated among contemporary timber dealers. Graciously, the trader donated a portion of his discovery to the Shanghai Museum so that all people might benefit from his find. Sometime afterward, however, the wood was revealed to be not *zǐtán* at all but an entirely new type of rosewood that most Chinese people had never seen before.

The wood, as it turned out, was rosewood from Madagascar. The clever trader had "discovered" this wood not in an imperial cache but during his worldly travels in search of new commercial supplies. Noticing its striking similarity to the famous *zǐtán* used in China centuries ago, he then passed it off as an imperial cache, exponentially increasing its value and maintaining the secret of his real discovery. In an attempt to save face, the museum renamed the wood "big-leaf" *zǐtán* (*dà yè zǐtán*)—despite its not even being of the same genus as the original Chinese wood.[1] This is how, according to commercial folklore, Malagasy rosewood received its Chinese name. This is also why rosewood from Madagascar has become so valuable: its coincidental likeness to one of the most coveted woods in imperial China.

This story of the Chinese "discovery" of Malagasy rosewood has the air of being centuries old rather than only a few decades. It is just one of a barrage of stories of the global "discoveries" China has made after the country's reform and opening in the 1980s and 1990s, the "going out" policy (*zǒu chūqù zhànlüè*) in 1999, and the first Ministerial Conference of the Forum on China–Africa Co-operation in 2000. From 1990 to 2013, Chinese foreign direct investment (FDI) stock in Africa grew from US$49 million to just less than US$25 billion.[2] By 2019, that figure was more than US$110 billion, increasing substantially as US FDI in Africa declined.[3] Trade dynamics show similar trends. In the early 2000s, as Africa's share of trade with Europe declined, the continent's trade with Asia increased by 18 percent per year, with China accounting for most of the growth.[4] By 2013, China was sub-Saharan Africa's largest trading and development partner, with trade valued at US$170 billion.[5] China "discovered" new resources abroad just as African countries "discovered" a new trading partner with particular qualities not shared by typical Western donors.

Most, but certainly not all, of these "discoveries" were of minerals and oil. China began importing coltan and tantalum from the Democratic Republic of the Congo (DRC) just as trade sanctions from the United States sent miners desperately looking for buyers.[6] China and the DRC also arranged one of the largest planned "resource-for-infrastructure" deals at the time (initially planned to be US$9 billion but ending up much less), exchanging a highway and railway for copper- and cobalt-mining rights in a loan setup that has since become characteristic of many Chinese investment projects in Africa.[7] Zimbabwe, another country facing trade sanctions by Western donors, began its Look East Policy and similar mining arrangements with China ensued.[8] Beyond minerals and oil, timber resources and agricultural products traded between China and Africa during the past two decades have seen steady growth.[9] Increased trade and investment is a result of growing bilateralism between Chinese and African national governments but also of emboldened Chinese entrepreneurs setting up mines, plantations, and export-import operations across the continent.[10] All of these connections represent experiments in new relationships between China and Africa. Many have failed—including, for the most part, China's initial resource-for-infrastructure deal with the DRC—but enough have succeeded over the past two decades to make China now Africa's largest trading partner and the go-to partner for big infrastructure projects that are so needed across the continent.

Accounts of burgeoning China-Africa relations demonstrate the newfound global frontier mentality China has inspired. First with the country's "going out" policy and now through the Belt and Road Initiative and the Asian Infrastructure Investment Bank, China increasingly turns to the global community to meet its development goals and resource needs, with Africa playing a central role. Soon after its inception in 2013, the Belt and Road was projected to be worth nearly US$1 trillion by its anticipated completion date of 2050, with 60 or so countries participating. Although this was remarkable at the time, the figures have only increased since. Now, with 140 countries already signed on, the most optimistic projections estimate the initiative to be worth more than US$8 trillion by 2050.[11] In either case, the Belt and Road is orders of magnitude larger than the post–World War II Global Marshall Plan—the only comparable investment plan boasted by Western leaders. Initially planned mostly within the geography of the historic Afro-Eurasian Silk Road but now including nearly all of Africa, much of Latin America, and a "polar Silk Road" component as well, this initiative has been referred to as "the new WTO," "globalization 2.0," "global commerce on China's terms," and a grand departure "from an America-centered world order."[12]

As global circuits realign and resettle to accommodate this new center of gravity, more and more resources are brought under China's reach. But this geographic expansion is only part of the story. As much as global media outlets frame "global China" as a tidal wave of new investment undulating outward or the rise of one superpower replacing another, it is rather the longer historical trajectory of China in the world that the country is itself invoking in using the term "the new Silk Road." In other words, there is a disconnect between what China as a global force means inside and outside of the country. Views on "global China" as they reverberate throughout Western media are incomplete. They in fact reproduce the same (mis)readings and (mis)judgments of Western views on China's economic rise in the early 2000s after it joined the World Trade Organization (WTO). In concluding this book, I reflect on these (mis)readings and offer an alternative interpretation of the deeper stakes of "global China"—namely, the country's decentering of Western-style globalization and consequent disruption of reigning Western orthodoxies that have heretofore been erroneously considered synonymous with the "global."

"discovering" global China

When speaking of "global China," scholars often speak quite literally of Chinese firms and Chinese people living and working outside the borders of China. Of course, the punctuated expansion of Chinese people and Chinese economies has been occurring as long as there has been a "China" to speak of. This expansion was first within the mainland itself as Han Chinese assimilated other cultures within their growing bureaucratic empire as early as 2000 BCE during the

Xia Dynasty. Global connections began to emerge much later, most prominently with the ancient Silk Road linking China to the rest of the Afro-Eurasian world starting in the second century BCE during the Han Dynasty. The early-fifteenth-century Ming Dynasty saw another moment of punctuated global expansion as the Yongle Emperor launched vast voyages overseas, led by the great Chinese explorer Zheng He. These maritime voyages reached across Asia to East Africa, establishing trade routes and collecting rare specimens of the world outside China. The Qing Dynasty saw similar excursions, as discussed further on the pages that follow. Yet it was the fall of the Qing, ending imperial China and beginning the country's occupation by Europe and Japan, that resulted in not just trade and exploration but also significant immigration.[13] The early twentieth century witnessed a wave of Chinese immigration to many parts of the world, including Africa. Unlike the immigration during the dynastic ages of exploration, this wave involved new Chinese diasporas established across the world. This was when, as noted in Chapter 2, Chinese labor was imported into Madagascar by the colonial regime to work on railroads and other infrastructure projects. Other African colonies experienced a similar influx of Chinese labor. And also as in Madagascar, many of these laborers did not return to their homeland after the work was finished. They established a life overseas, growing cash crops or establishing small-scale merchant operations.[14]

This wave of immigration laid the groundwork for today's global projects, such as the Belt and Road Initiative. Although China has been a global force for centuries—millennia, in fact—the coordination and scale of the country's geographic expansion, primarily through state and private capital, are unprecedented in the current century. This twenty-first-century unfolding is what scholars and the media often invoke when they refer to global China. The unfolding began in the 1990s as the country demonstrated unprecedented economic growth. In 2001 China joined the WTO, the highest signification of integration into the global economic system possible at the time. It was a pivotal moment for China both domestically and internationally, yet—much like the concept of global China in general—it signified two very different things for those inside the country and those outside of it.

After fifteen years of diplomatic struggle, economic reforms, and lengthy negotiations with the United States and Europe, China's formal membership in the WTO was "an event of historic proportions for the world trading system," as told by Secretary-General of the United Nations Kofi Annan, and a "historic moment in China's reform and opening" as told by the *People's Daily* in China.[15] Yet even though the importance of the event was agreed upon across the globe, its exact significance (particularly for the future of globalization) was contested.

In the West, there was little doubt that political and social reforms in China would soon follow the economic. "Let China get rich," contemporary artist Ai Weiwei summarized the American perspective in retrospect, "and then watch as freedom and democracy evolve as byproducts of capitalist development."[16] George H. W. Bush's famous remarks echo this sentiment exactly. "No nation on Earth," Bush declared, "has discovered a way to import the world's goods and services while stopping foreign ideas at the border. Just as the democratic idea has transformed nations on every continent, so, too, change will inevitably come to China."[17]

Change, no doubt, came to China, but not in the form of Western-style democratic reforms. The Communist Party now maintains as tight a grip as ever on business activity, monetary flows, and public opinion. Most recently, China is at the forefront of experiments to control mega-corporations and their leaders, such as Jack Ma's fintech giant Ant Group—the company behind China's largest private-sector tree-planting campaign, Ant Forest, discussed in Chapter 6. Some Western pundits argue that such a heavy-handed state will surely be "the undoing of China's economic miracle," but others acknowledge the long tradition of Chinese state intervention in both markets and society, even during the lauded Deng-era reforms.[18] Deng Xiaoping allowed for zones of market experimentation that eventually led to overwhelming growth, as noted in Chapter 1, but his administration nonetheless maintained far more control of the economy than any Western country would dream. Moreover, while opening the economy, Deng and his administration took great care to avoid the "ideological weeds," "cultural poisons," and "spiritual poisons" of liberal ideologies based on individual freedoms.[19] It was economic development and a rise in living standards that Chinese leadership was in search of, not social or cultural change. Experimentation with market logic was tolerable only so long as it was highly controlled and did not threaten party rule or social stability. To this day, it is not clear whether the Communist Party's heavy-handed state was and will be the "doing" or "undoing" of China's economic miracle. But it *is* clear that most people in China at the time of its WTO accession had no such expectations of a liberal transformation as was widely believed in the United States. Instead, they came to "the very logical conclusion that American values are fine for America but would never work in the Chinese system."[20]

When China opened its economy to the world, the country did not democratize, liberalize, or Westernize. This much is now largely agreed upon by Chinese and Westerners alike. Now that China is expanding its economy *out into* the world, the same should be expected. That is, Westerners should *not* expect a renewed round of Western-style globalization—even less, Western-style neocolonialism. In understanding China's global ambitions in the present moment, it is imperative that we do not reproduce our misreadings of the past. To this end, it is helpful to explore China's current global engagements and better understand some of the ways that they differ from Western-style globalization as it has proceeded in the late twentieth and early twenty-first centuries.

Refuting analogies of twenty-first-century global China with Western colonial powers of centuries prior, ethnographer Ching Kwan Lee declares, "There is no military occupation by China in Africa, no chartered companies with exclusive sovereign trading rights, no religious proselytizing."[21] Even when comparing Chinese state investments (circulating both within and outside the country) with contemporary global capital circulating via corporations, Lee maintains, differences are manifold. These differences are partly why many economists misunderstand Chinese policies and planning as "inefficient." From the perspective of pure profit maximization, Chinese capital may indeed be "inefficient," but from the perspective of long-term growth and social stability, efficiency (or lack thereof) is secondary. As noted, all goals pursued by the Chinese Communist Party—economic or environmental—are subordinated to the paramount goal of maintaining political power, which largely rests on maintaining social stability and public approval. This applies *globally*, when it comes to China's overseas investments, as well as domestically. Global China, then, is driven not by capitalist logic but by the logic of a state selectively deploying capitalist tools (among a medley of others) to maintain a strong and prosperous country within the global arena.

As Lee demonstrates through the case of Chinese mining investments in Zambia, China's focus beyond the economic has created negotiating room for recipient countries. "Exactly because of its more ambitious agenda, which cannot be reduced to profit," she notes, "Chinese state capital has been more concessionary and negotiable with Zambian state and society than global private capital."[22] Chinese state-owned enterprises are interested not only in securing minerals at the right price but also in maintaining good diplomatic relations. This has the effect of localizing Chinese state capital more within the host country, Lee notes, and even potentially empowering the host country as a counteragent. Chinese state capital is thus "sticky" and place-bound, "opening more room for bargaining" compared with finicky and price-sensitive global private capital.[23]

Beyond the unique "stickiness" of Chinese state capital, China's bilateral and multilateral relations offer other clear examples of a non-Western approach, most notably in terms of the diplomatic principle of nonintervention. Nonintervention in other countries' domestic affairs has been a pillar of Chinese foreign diplomacy since the establishment of the People's Republic of China in 1949, when the Five Principles of Peaceful Coexistence were included in the country's constitution. These principles include mutual respect for each other's sovereignty and territorial integrity, mutual nonaggression, mutual noninterference in each other's internal affairs, equality and mutual benefit, and peaceful coexistence. Following these principles, China has been cautious about authorizing sanctions and military interventions—typical Western responses to countries, such as the DRC, Zimbabwe, or Zambia, that demonstrate political misconduct or human rights abuses. Western sanctions against African countries such as these in the 1990s and early 2000s created an economic vacuum that in some cases, as discussed earlier, China ended up filling.

In addition to noninterference in cases of political misconduct or human rights abuses, China has been notoriously opposed to conditionalities on loans and investment that require specific partner country reforms—the conventional model of Western loans and aid coming from the World Bank and the International Monetary Fund during the Washington Consensus era. Chinese jurists have gone so far as to label such intervention on the part of Western powers an imperialistic strategy to interfere with and exert undue influence in domestic affairs.[24] Certainly, China's strategy of nonintervention is a direct response to the history of colonization the country has itself experienced. The sentiment survives to this day, invoked during recent high-level talks between the United States and China in which, for example, Chinese representative Yang Jiechi rebuked the United States for "long-arm jurisdiction and suppression" as well as the use of "force or financial hegemony" to promote American interests abroad. The text of the Belt and Road Initiative clearly advocates for a noninterventionist approach, cautioning against domination and superiority.[25] Whether true in practice or not, the rhetoric and performativity concerning nonintervention and "win-win" partnership goes a long way in establishing goodwill from member countries.[26] As African policy analyst Gyude Moore concludes, "China wins *a lot* more . . . but at least [and noting a few exceptions] Africa doesn't *lose*."[27]

From a Western perspective, China's principle of noninterference is a convenient excuse to wreak havoc at the global level, propping up fascist regimes and extracting resources with impunity. Moreover, it impedes the "natural" march of progress toward liberalization and democracy that is so often assumed. The gesture toward "self-determination" China offers its partners is cast aside as rhetoric, if mentioned at all. This can be witnessed in, for example, Hilary Clinton's advice for African countries to be wary of others that "come in, take out natural resources, pay off leaders and leave" or Mike Pompeo's more explicit accusation of China's extension of "predatory loans" to African states in order to extract profits and offer little in return.[28] More recently, former White House national security adviser Robert O'Brien criticized China's presence in Africa, asserting that "whereas Beijing promotes a journey to China dependence, the U.S. promotes a journey to self-reliance."[29] The irony of Western advisers criticizing China while bolstering their own efforts abroad did not go unnoticed by the African countries caught in the middle.

In terms of the environment, China's principles of nonintervention also face critique. Critics are quick to point out that even as China ratchets up environmental standards domestically and makes local and provincial leadership contingent on meeting environmental targets, the country nonetheless continues to invest in overseas projects that do not meet its own standards. The proposed fix is to require that China's overseas projects be held to the same environmental standards as the domestic ones.[30] Yet this seemingly logical and transparent solution appears as such only in the eyes of countries that follow a long tradition of imposing conditionalities on development assistance. In light of the Chinese diplomatic tradition of nonintervention, the prospect is not so simple. When a country becomes a member of the Belt and Road Initiative, that country is establishing, so the rhetoric goes, a mutually beneficial win-win partnership. Whether or not this is actually the case is up for debate, but Chinese leadership is very careful not to commence the partnership with a list of conditions. Partner countries instead are offered a suite of potential projects ranging in environmental impact—dirty coal, clean coal, natural gas, renewables—and it is up to the partners to decide what works best for their country. China has grown at an unprecedented rate, and the types of projects and environmental standards suitable for the Chinese context may not be desirable from the host country's perspective. Chinese investors are certainly not in a position to determine what best suits their partners' needs.

Thus, China's approach to foreign diplomacy can be seen on one hand as a convenient cop-out allowing China to wreak environmental havoc and indirectly prop up authoritarian dictators or, on the other hand, as an unnegotiable tenet that resists colonial tendencies and distinguishes China from the pack of Western donors. Western media hum primarily to the tune of the former, and Chinese media favor the latter. The deeper point is that China and the West (again, especially the United States) maintain different approaches to intervening in the world based on different underlying assumptions. Whereas Western interventions are influenced by a cultural tradition of universal humanism inspired by visions of progress and liberalization for all the world's people, China's approach acknowledges that different systems work for different people. This approach translates into noninterventionist foreign diplomacy and bilateral loans that generally lack Western-style austerity measures or governance reforms.

To summarize, the Western surprise over China's failure to socially or politically liberalize in the early aughts is indicative of a misreading of China's domestic ambitions. Similarly, contemporary debates over China in the global arena that cast the country as a neocolonial actor or a new hegemonic force are indicative of an analogous misreading of China's global ambitions. In this misreading, globalization as understood through a Western lens is superimposed on a new Chinese superpower. China is slotted into a hegemonic space that all globalizing forces assume, and the rest proceeds as expected: resource acquisition and exploitation, predatory loans, self-serving infrastructure projects masquerading as aid or assistance. This I refer to as the Western "discovery" of global China—discovery being a moment in which a newfound empirical reality is absorbed and explained through the reigning conceptual apparatuses of the discoverers.

a new paradigm

My analysis of China and its worldly encounters in this book offers a different approach to understanding the country's global ambitions. I argue that the concept of "global China" is not simply extended geographic presence or intensified capital investment but rather the emergence of a new orientation—or "reorientation," as economic historian Andre Gunder Frank would phrase it—on the global scene.[31] More precisely, and as elaborated later in this conclusion, global China disrupts the long-standing nineteenth- and twentieth-century and early twenty-first-century conflation of "Western" and "global."

This alternative reading of global China has both economic and cultural ramifications. Economically, the story is not all that different from what I have discussed here. China's export-led growth of the 1990s yielded to investment-led growth in the early 2000s, much of which included infrastructure—bridges, high-speed rails, dams, trees, and buildings, as far as the eye can see. As domestic investment avenues became saturated—no more high-speed rails to build, no new urban skylines to erect—China turned outward. Rather than squandering this newfound expertise and capacity, the country is investing overseas. Domestic investment-led growth that was headed for stagnation is being reinvigorated by global investment-led growth. As Lee nicely summarizes, "the phenomenon of global China is about China seeking spatial and political fixes to its resource and profit bottleneck, in the context of a national and global overaccumulation crisis, with no preordained or guaranteed outcome."[32] Indeed, the booming rosewood market demonstrates this clearly, with too much "hot money" and not enough productive outlets for growth. Hence, rosewood and other endangered species and cultural items are reinvented as new speculative investments.

This economic understanding of global China is not especially revelatory. Similarly to China's domestic economic opening, it clearly serves a strategic purpose that can be understood by Chinese and Westerners alike. The difference, however, lies in the reality that for China (and perhaps many countries that cannot be considered fully neoliberal in the Anglo-American sense) the economic is always subordinated to the political, with economic growth never a goal in itself but rather a means to a political end. Special economic zones in the 1990s, for example, were considered "a laboratory in which to experiment with capitalist principles in the advancement of socialism" and always retained deep-seated state involvement.[33] China's economic "miracle" was a strategically planned route to party support in a post-Mao era that retained no cult of personality ensuring approval. The Belt and Road Initiative provides a new engine for growth but also a new global legitimacy that ties the Chinese Communist Party to something much bigger than delivering growth in gross domestic product. It ties the party to global Chinas of the past and global dynamics of the future.

Culturally and symbolically, the concept of global China, and especially the "new Silk Road," alludes to a deeper historical continuity, connecting global Chinas of past and present. It emphasizes Silk Roads old and new while pushing the interstitial period of Euro-American hegemony into the background. In this way, global China disrupts the narrative of globalization commonly recited. For the past two centuries at least, "global" and "cosmopolitan" have simply meant "Western." Seemingly global forms—capitalism, science, the international system of nation-states—are in fact decidedly Western ideas that have gained global salience. The "Western" is at present indistinguishable from the "global"—so much so that few Westerners even recognize the tacit conflation. We think of "Bretton Woods" institutions, for example, as purely "global" institutions but forget their provincial origins in rural Bretton Woods, New Hampshire—a small resort town in East Coast America during an unprecedented postwar period when the United States was miraculously able to unilaterally set the global agenda. The conflation, of course, goes far beyond Bretton Woods. Western concepts across the board, as we have seen, are too often presented in universal terms. The case of nature preservation—a peculiarly Western approach to understanding the human place in the world that has been exported across the globe as universal science—demonstrates this precisely. But the tendency to conflate the Western and the global is fast changing, and "global China" is that vast assemblage of forces hastening change.

Global China, in other words, disrupts contemporary narratives of Western globalization and, in so doing, revitalizes the much longer historical narrative of globalization in the classical sense of the Silk Road. Glossing over Euro-American hegemony of centuries prior while alluding to a deeper historical Chinese continuity, the "new Silk Road" unites global Chinas of past and present. A space long considered synonymous with Western hegemony ("the global") has been reclaimed in the name of an idealized past via Chinese leadership in the present. As President Xi Jinping historically announced when inaugurating the Belt and Road Initiative, "exchange will replace estrangement, mutual learning will replace clashes, and co-existence will replace a sense of superiority."[34] All of this will occur through state and private Chinese capital flows that hearken back to ancient times. China thus emerges as a global force while fortifying the cultural and political identity of its people.

The result is a new orientation in which ideas of China—as well as Chinese ideas—once again orient the "global" in a fundamental way. Chinese worldly encounters order global imaginaries as much as global realities and thus count as "discoveries." The clever Chinese trader who passed off his newfound rosewood for an imperial cache, discussed earlier, did not "discover" Malagasy rosewood any more than Columbus "discovered" America. Yet, somehow, when these two explorers encountered new lands and resources, both of their encounters counted as discoveries. Having such encounters count as discoveries is in many ways more important than the discovery itself. The significance of the concept of global China, then, is not so much the increasing acquisition of resources scattered across the globe as the ability to "discover" and redefine these resources at the global level.

global Chinas of the past

China's "discovery" of Malagasy rosewood in the 1980s and 1990s was of course not the first—not even for China. Rosewood has in fact been traded between China and Madagascar for centuries, as noted in the introduction to this book. Since Zheng He's voyages to Africa during the early Ming Dynasty, rosewood has been considered a special gift sent to China from Madagascar.[35] During the mid-Qing, after further endangerment, the imperial court sent explorers in search of new rosewood supplies throughout Asia, Africa, and South America. Rosewood from Madagascar was one of many rosewood varieties that made it back to China. Looking beyond the obvious, China's dynastic search for rosewood was much more than a search for physical wood. Enrolled in this global search for rosewood was a search for global identity. For the ruling elite, this meant not only gaining physical access to far-off resources but also having a hand in reframing them and redefining their import at the global level. It meant not only being a part of global flows but also reshaping how they were understood and experienced. It meant, in short, transforming a mere encounter into a "discovery."

Consider, for example, the work of Lang Shining (1688–1766)—an Italian Jesuit missionary and painter who lived and worked in China. Known more commonly by his given name, Giuseppe Castiglione, Lang was one of the highest artists commissioned by the emperor during the Qing's cultural and economic zenith—another era in which one might also speak of a "global China." One of Emperor Qianlong's favored artists, Lang represented in his work a unique blend of Western and Chinese traditions.[36] He was sent on expeditions to paint exotic animals "discovered" across the world. Lang's travels to Madagascar resulted in his famous *Cochin Lemur,* which likely constituted the only work of art in the world at the time that combined elements originating from Europe, China, and Madagascar in a single canvas (Figure c.1).[37]

In commissioning Lang to depict this and other exotic species across the globe, Emperor Qianlong was not simply attempting to learn what these species looked like. Rather, he was having a say in their worldly representation. That is, he was establishing some degree of cultural hegemony over representations of the global "exotic." Surely, the emperor knew very well that the ability to iconically depict exotic species in faraway places can be far more powerful than faithfully representing the physicality of these species. *Cochin Lemur* is not an act of faithful representation but rather a strategic making of the global. Lang's work borrows from Western representational forms in order to give greater legitimacy to the traditional Chinese backdrop. A new biological specimen foregrounded through Enlightenment realism contrasts with a background of classical Chinese blossoms, mountains, and trees, representing a beautiful clash of divergent artistic traditions, all funded by the Qing emperor. By commissioning artistic renderings of worldly encounters in this fashion, the emperor was representing the "exotic" in a style that could be recognized as both global and Chinese. He was, in other words, securing a place for China on the world stage. The piece resoundingly announces the global "discovery" (not mere encounter) of an exotic species through a distinctly Chinese register and, in doing so, reads almost as a manifesto of the "global China" of that particular era.

The same is true for contemporary global China. More than physically controlling resources, the greater challenge is to control their representation and reception at the global level. Indeed, it is through such strategic representations

Figure c.1. *Cochin Lemur* by Lang Shining (Giuseppe Castiglione), painted in 1761 for Emperor Qianlong. National Palace Museum, Taipei, Taiwan / Wikimedia Commons.

that mere encounters are consecrated as "discoveries." For China in particular, establishing a global presence—or being able to frame and reframe the global—is not only about looking forward but also about looking back. Traditional narratives of China as a global powerhouse play a prominent role in the country's contemporary global ambitions, including most prominently the Belt and Road Initiative—China's "modern-day Silk Road." This global project is not simply an outward expansion aimed at the acquisition of land and resources from abroad. It is even less an initiative aimed at inculcating Chinese values in other governments or populations. But it does aim to disrupt the past centuries' tradition of Western ideas orienting global spaces. This unprecedented global initiative redefines a space that could be considered at odds with traditional Chinese culture—the space of contemporary globalization—in terms that now strongly resonate with traditional Chinese themes.

looking forward and backward

Reconfiguring global spaces in terms that resonate with Chinese interests can be witnessed on a much smaller scale in the installation of traditionally styled rosewood furniture in the modern Chinese home. Traditional rosewood furniture is—unsurprisingly—designed for the *traditional* Chinese home. The main hall in the traditional Chinese home, where elaborate rosewood furniture sets are most prominently featured, is for formal display. It borders on a public space, where visitors are received and a family's fine tastes are accented. Rosewood furniture—dense and angular—is designed not for lounging but rather for filling this deeply symbolic space with the first and most prominent demonstration of familial power.

The *modern* Chinese home, in contrast, is problematically Western. The traditional main hall has been usurped by the living room—a space of relaxation rather than a formal space of display. This Western layout, which has become the global model for contemporary domestic life, likely retains a degree of unfamiliarity when it comes to China's older generation. From this perspective, the installation of traditional Chinese furniture in the modern Chinese home provides a way to take back that space—to imbue meaning and familiarity in an otherwise sterile arrangement. At a larger scale, the popularity of rosewood furniture provides an appeal to traditional values in the face of rapid modernization, industrialization, and now already in some places post-industrialization. Amid China's rapid transformation, rosewood furniture demonstrates not only the wealth and status of the family but also an analogous transformation of a

Western space into something more familiar and habitable. As with the Belt and Road Initiative but on a much smaller scale, rosewood furniture provides the ability to reshape foreign spaces in a way that is distinctly Chinese. It allows Chinese people to double down on their cultural identity rather than lose it amid such rapid economic growth. It enables both a move forward and a step back.

It is this simultaneous step forward and back that most succinctly characterizes the phenomenon of "global China": a spreading out while simultaneously drawing inward, an agreement to join the global community but also to remain steadfastly Chinese. This paradoxical nexus of facing outward and inward can also be seen, by way of another example, in Deng Xiaoping's "Socialism with Chinese Characteristics" or, more recently, Xi Jinping's "Socialism with Chinese Characteristics for a New Era." Traditional Chinese values are not lost in these doctrines but repurposed and transformed.

To be properly understood as a global force, then, both the "global" and the "China" must be considered together—the simultaneous look forward and backward, expanding yet drawing in. Chinese art, again, is helpful in teasing out this hidden paradox. In fundamental ways, contemporary Chinese art mirrors the work of Lang Shining's *Cochin Lemur* from the Qing Dynasty. Drawing on both Chinese traditional culture and twentieth-century Western art forms, contemporary Chinese art requires a basic fluency in both to be fully understood. The work of Ai Weiwei, one of China's most famous contemporary artists (and commenter on divergent Chinese and American cultural perspectives, as quoted earlier), demonstrates this most clearly. Take, for example, Ai's *Moon Chest* (Figure c.2), constructed of one of the rarest rosewoods, which according to traditional Chinese methods requires no jointing or joinery. These chests are iconic symbols instantly recognizable to anyone remotely familiar with traditional Chinese culture. But as contemporary art, they have been punctured by large holes, rendering them effectively useless and the traditional icon disfigured, except when viewed from a certain angle, at which point another powerful Chinese icon emerges—the phases of a lunar eclipse.

In true paradoxical fashion, this art looks both forward and backward, considering traditional Chinese icons, Western artistic conventions, and contemporary capitalist dynamics, all at the same time. Each value subverts the other while simultaneously relying on its meaning. As with Lang Shining's *Cochin Lemur,* disparate elements are forced to occupy the same incommensurate space—East, West, Rest. The result is a prismatic effect as the viewer slides from one side to another, forcing the synthesis of multiple artistic traditions in a single piece. In Ai's *Moon Chest,* traditional Chinese craftsmanship is effectively destroyed by negative holes that are then, through a Western twentieth-century artistic ma-

Figure c.2. *Moon Chest* by contemporary Chinese artist Ai Weiwei on display at the *According to What?* exhibition at the Brooklyn Museum, 2014. One of seven chests carved of *huáng huālí* rosewood with a hole puncturing the center, revealing the cycles of a lunar eclipse. © George Zhu.

Figure c.3. *Chair with Two Legs on the Wall* by contemporary Chinese artist Ai Weiwei on display at the *According to What?* exhibition at the Brooklyn Museum, 2014. © George Zhu.

neuver that turns absences into images, transformed once more into yet another Chinese cultural symbol—the moon. In the disfiguration of these cultural values—holes in traditional rosewood chests or the warping of traditional rosewood chairs (Figure c.3)—a new value emerges. This is the object's value as a piece of contemporary art, now worth exponentially more than the already hyperinflated rosewood material with which it was made. Thus, the viewer is forced to see all of these values—cultural values, commodity values, resource values—juxtaposed in a single conceptual piece, readable in different ways by different people.

The juxtaposition is not seamless but rather anxious and fragile. Much like dynastic rosewood furniture occupying the modern Chinese home or the traditional icon of the Silk Road occupying the present-day global arena, contemporary Chinese art highlights a certain rupture that comes along with disparate elements inhabiting the same space. As anthropologist and assemblage scholar Aihwa Ong rightly observes, contemporary art is "a distinctive mode of space rupturing and conceptual reconfiguration."[38] It bears striking similarity to the rebranding of globalization in ancient Silk Road terms or the installation of traditional rosewood furniture in the modern Chinese home. Each reveals the discontinuity and paradox associated with looking both forward and backward. Flat-screen televisions and mountains of toys awkwardly crowd traditional rosewood furniture sets as they adorn the modern Chinese living room. Amid an array of electronics and disposables from another era, there is a certain ridiculousness that comes along with sitting in such fanciful chairs: thrones repurposed for the modern consumer. As much as dynastic rosewood furniture may reclaim the Western living room through a distinctly Chinese register, it nonetheless sits awkwardly in a space designed for modern lounging.

The uselessness of a Ming Dynasty throne in one's modern-day living room is comparable to the uselessness of Ai's distorted tables or punctured moon chests. They both underscore a certain incommensurability—a haphazard throwing together of elements that would otherwise never cohere. In contrast to Lang's *Cochin Lemur,* which strives to portray a seamless amalgamation of East, West, and Rest, ignoring the fragile juxtaposition that dominates the canvas, Ai's rosewood cabinet and chairs instead *emphasize* that rupture and uncertainty. Whereas *Cochin Lemur* uses the canvas to conjure a unified global terrain consonant with Chinese interests, ultimately pointing to the incontrovertibility of the global China of that time, *Moon Chest,* in contrast, forces one to consider the absurdity of traditional Chinese craftsmanship circulating in a global capitalist market—the incongruity, that is, of traditional Chinese values on the global scene. The former glosses over a paradoxical space of global juxtaposition; the latter accentuates it. The former is a forceful assertion of global China, whereas the latter is a critique. The Belt and Road, in this sense, jives more with the former, emphasizing a united front over the unruly assemblage of countries of which it is composed. Yet, even if obscured, the fragility and tensions remain.

The concept of global China—as with contemporary Chinese art, as with rosewood—forces one to confront the paradoxical space of the global. This space is not a seamless "world system" where—uneven as it may be—there is a greater totality to speak of, in which one power usurps the place of another. It is rather, as Ong notes, a moment of "space rupturing and conceptual reconfiguration" wherein the viewer is forced to consider not one side or the other—global / situated, forward / backward, traditional / modern—but both in tandem. Here, one finds echoes of rosewood's paradoxical space of contestation and common ground: the meeting point of tree and forest, when the first blows of an axe reveal the bright red heartwood underneath and the value of rosewood as pristine old growth collapses into the value of the tree as a vibrant historic wood. Global China also reveals such tensions and sits amid such contestation.

Western interpreters of China's global ambitions should take note. Global China should not be interpreted simply as Chinese investment reshaping the future of global capitalism or Chinese people assuming the cultural hegemony of the West as it moves forward. It should be seen not as a shifting center of gravity from one side of the globe to the other but rather as a new source of global rupture. Global China is a disruption of the past two centuries' tacit conflation of "Western" and "global."

repatriating endangered species conservation

In 2020, Madagascar's minister of environment and sustainable development, Baomiavotse Vahinala Raharinirina, boldly announced that "the conservation of our biodiversity through Madagascar's protected areas' system for thirty years was a failure."[39] "We have to change the paradigm," she continued, "and to move toward a system which doesn't exclude humans and [doesn't] put local communities on the sidelines; it should be deeply social." Of Madagascar's 144 protected areas, only 15 are directly managed by the ministry, and an additional 46 are managed by the quasi-governmental organization Madagascar National Parks. The rest are managed by domestic and foreign NGOs supported by international donors. This is the transnationalization of Malagasy lands discussed in Chapter 4, the "inverted commons" wherein Madagascar's biological wealth is placed largely in Western hands.[40] "It is not fair," Minister Raharinirina solemnly proclaimed, "to see small villages near the country's most important tourist destinations, where people continue to live in darkness and [with] poor drinking water."

Around the same time, another seemingly unrelated shift was occurring in another part of the world. In response to the past decade's art "thefts"—those discussed in the introduction to this book—national museums across Europe began repatriating especially controversial Chinese artifacts unprompted. Practically, this repatriation reduces the threat of future targeting for thefts and offers the chance to improve diplomatic relations with China. Conceptually, it decenters a Western perspective on legality and legitimates a non-Western claim to these resources. One could imagine a similar type of repatriation of conservation efforts—repatriation in this sense meaning the act of "returning" to non-Western peoples the ability to define their own legitimacy and legality in regard to their environment. This kind of repatriation would change how conservation is approached and executed so that the goals are not determined at the outset and so that the very question of preserving natural environments is put on the table for revision. Such an approach to conservation would entail acknowledging diverse histories that come to bear on contemporary environmental efforts and entertaining new perspectives on a sustainable future. This could mark a first step toward truly collaborative and novel interventions that go beyond strict wilderness preservation to envision convivial futures of human-environment transformation.[41] In a globalized world in which the West and its "others" will increasingly encounter one another as equals, this seems like an appropriate way forward for endangered species conservation.

Where does China fit in this process of environmental repatriation? It is one thing to empower African nations harboring some of the world's last remaining species and quite another to stand by as China consumes the lot. Aspiring to liberal notions of plurality and multiculturalism—as repatriation often does—presents no cognitive dissonance when it comes to indigenous peoples and local communities, especially when these people are not in a position to change the conservation agenda. But when it comes to a powerful and largely illiberal China, the rhetoric breaks down; the appeal is lost. Even worse, China's divergent approach to endangered species conservation and the environment—engaging in captive breeding and large-scale ecological engineering, planting trees on a large scale rather than protecting them in nature—seems only to embolden African countries to stand apart from the West, such as the Southern African Development Community vis-à-vis CITES (see Chapter 5). China offers an alternative model for environmentalism based on neither the preservationism of Western conservation nor the conditionalities of Western donors. It is an appealing invitation to self-determination that other countries wishing to develop their resources as well as conserve them will likely pursue—perhaps to the detriment of global sustainability, certainly to the detriment of pristine "nature." Given this potential hijacking of the Western conservation agenda by a rising global China, it may appear that the only solution is to double down and remain true to one's own vision of a sustainable future and represent an environmental force for good in a conflicted world.

Yet, authoritarian, democratic, or otherwise, there is no solving the world's biodiversity crisis without working with China. Indeed, no global environmental crisis, including the looming threat of climate change, can be addressed without collaboration with China. This may be the biggest "pill," so to speak, for Western environmental movements to swallow. As stories of Xinjiang, Hong Kong, trade sanctions, and a long list of other highly divisive issues fill the news, there seems to be no bridging the insurmountable political and cultural gap between China and the West. From one perspective, there can be no tolerance for collaborating with an authoritarian regime; from the other perspective, there can be no tolerance for subjecting one's country to foreign imperialist ideologies.

The challenge of collaboration, however, is not about clichéd ideals of embracing cultural difference; it is about realizing the practicalities of global compromise and working together to solve collective problems no matter how ideologically opposed the two sides might find themselves. This type of collective planetary living is neither romantic nor ideal and should not make one feel particularly warm or fuzzy. It requires "staying with the trouble" and "making odd kin," as critical theorist Donna Haraway calls it, or "making do" in "the ruin that has become our collective home," as ethnographer Anna Tsing calls it.[42] Preserving untouched landscapes devoid of humans—a peculiar aspiration not shared by most people on this planet—seems counter to such convivial life and planetary collaboration.

What collective future, then, might we aspire to? And what will become of Africa and other seemingly out-of-the-way places as the decentering of the West advances and China's global rise continues?

A few months before Minister Raharinirina's ardent speech, Madagascar's president, Andry Rajoelina, started a campaign to plant sixty million trees, as he noted, "so our descendants can live healthy lives."[43] Sixty years of independence from colonial powers in 2020 and sixty million trees to celebrate, to "reforest for the benefit of future generations": this is by far Madagascar's largest tree-planting campaign to date. Nearly two hundred thousand seedlings were gathered for the launch, and a total of one million trees were planted over five hundred hectares, including fast-growing exotics, fruit trees, and endemics. Five tons of "seed balls" were also dispersed by drones and airplanes, with an expected survival rate of 60 percent.[44]

Madagascar's tree-planting drive continues to be faulted: for the use of nonnative species, the lack of a clear plan of exactly where to reforest, and overstretched NGO staff not willing to sacrifice protected areas management for planting trees. Similar critiques will likely arise when it comes to the implementation of Madagascar's new anthropocentric approach to conservation, as envisioned by Minister Raharinirina. Let us remember amid these critiques that Madagascar's vision of a sustainable future is different from the West's and different, for that matter, from China's. Imposing visions of either Edenic landscapes or ecological civilizations on African environments is likely only to exacerbate global tensions. A better model for planetary collaboration relies not on utopian aspirations of preserving pristine nature nor the techno-optimism that comes along with engineering civilizations anew but rather on modest suggestions and support for living and growing amid our collective ruins.

notes

acknowledgments

index

notes

introduction

1. "Break In at Stockholm Royal Palace," *The Local,* August 6, 2010, https://www.thelocal.se/20100806/28220.
2. Indeed, a few of these stolen items have allegedly resurfaced in Chinese museums. Reporters have even questioned whether the Chinese government might be behind the thefts. Whether or not the state is in fact involved, given that China has been outspoken about the need to repatriate pillaged art and consolidate Chinese cultural artifacts within its national borders, it is unlikely that the government would judge harshly those who end up with some of these repatriated artifacts. Alex W. Palmer, "The Great Chinese Art Heist," *GQ,* August 16, 2018.
3. Rosewood, as it translates in Mandarin (*hóng mù*), is an informal term for a group of hardwood species that are usually red in color and are widely used for furniture making in China. Original species in this group include *huáng huālí* (typically represented by *Dalbergia odorifera*) and *zĭtán* (typically represented by *Pterocarpus santalinus*). These species have since become virtually extinct, and logs that either remain as trees or have been cut and carved into furniture can be, if of a certain age, more valuable than gold. These species have largely been replaced by other species of the *Dalbergia* and *Pterocarpus* genera. Rosewood from Madagascar (typically represented by *Dalbergia louvellii,* among others) is one of these replacements. It is in fact the most valuable of all replacement species, worth up to US$40,000 per cubic meter. Huang Wenbin and Sun Xiufang, "Tropical Hardwood Flows in China: Case Studies of Rosewood and Okoumé" (Washington, DC: Forest Trends Association, 2013).
4. United Nations Office on Drugs and Crime, *World Wildlife Crime Report* (New York: United Nations, 2020), Figures 3 and 4.
5. Ibid., Figure 3.

6. Ibid.
7. Yifang Cui, "Speaking for the Chicken Cup: A Case Study in Chinese Art Collecting" (master's thesis, Australian National University, 2017).
8. John F. Burns, "Qing Dynasty Relic Yields Record Price at Auction," *New York Times,* November 12, 2010.
9. Palmer, "Great Chinese Art Heist."
10. In this book, I use "conservation" and "preservation" interchangeably, typically referring to both as largely Western. In their historical usage the two terms have been opposed (as reviewed in Chapter 6), with "conservation" referring to the more utilitarian aim of conserving natural resources for future use and "preservation" referring to protecting nature for its own sake as "untouched" wilderness. Yet by the end of the twentieth century, particularly with the emergence of the field of conservation biology and its tight relations with the protected area (largely "untouched nature") approach, conservation was often conflated with preservation. Most conservation efforts in the tropics today retain a preservationist approach of cordoning off an area of land for nonuse. This is frequently combined with other areas of land that are repurposed for sustainable use (and thus recall conservation in the more traditional sense). Overall, however, the term "conservation" as used today in both rhetoric and reality is deeply shaped by a preservationist mentality, which, I argue, is particularly Western (see Chapter 6 for further discussion).
11. The most forceful example of this, discussed further in Chapter 5, is the threat to leave the Convention on International Trade in Endangered Species of Wild Fauna and Flora (CITES) leveled by southern African countries. This has been escalating since the CITES meeting in 2016. Most recently, in June 2021, tourism ministers from Zimbabwe, Namibia, Zambia, and Botswana (countries that contain more than half of the remaining wild elephant population) met to consider establishing an alternative international approach to CITES. Ray Ndlovu, "Where Elephants Roam, Leaders Float Idea of Reviving Ivory Trade," *Bloomberg,* June 18, 2021. The claim is that CITES has "unfortunately been hijacked by the anti-international wild trade Western animal rights groups who also fund it." Emmanuel Koro, "SADC to Trade in Ivory without UN Wild Trade Agency Control," *Zimbabwe Independent,* June 25, 2021.
12. See, for example, John Zinda, Yifei Li, and John Chung-En Liu, "China's Summons for Environmental Sociology," *Current Sociology* 66, no. 6 (2018): 867–885.
13. United Nations Office on Drugs and Crime, *World Wildlife Crime Report.* Figure 7 of this report shows that approximately 60 percent of rosewood seizures during this time were of rosewood from Madagascar.
14. This observation has been made elsewhere as well. Eva Keller, *Beyond the Lens of Conservation: Malagasy and Swiss Imaginations of One Another* (New York: Berghahn Books, 2015); Andrew Walsh, "The Obvious Aspects of Ecological Underprivilege in Ankarana, Northern Madagascar," *American Anthropologist* 107, no. 4 (2005): 654–665.
15. Significantly more money, however, has been spent on rosewood logging, with that figure alone reaching the equivalent of more than US$1 billion. Conservation finance, including provisions that specifically mention rosewood or are intended to curb rosewood logging, remains in the hundreds of millions.
16. One reporter quoted a close collaborator of rosewood traffickers: "We hear arrogantly from Chinese buyers, that the export will always continue in one way or another, because with their money, they think they can go through even the most highly placed doors" (translated

from French). "La 'bolabolacratie,'" *Madagascar Tribune,* March 1, 2012. Fighting these elite networks, a special task force of 120 troops and a helicopter was established to secure protected areas in northeastern Madagascar from rosewood logging. Funding was provided by international donors (the World Bank, USAID, the United Nations Educational, Scientific and Cultural Organization [UNESCO], and others), and conservation NGOs (the World Wildlife Fund, the Wildlife Conservation Society, and others) provided logistic and material support. Cynthia Ratsimbazafy, D. Newton, and S. Ringuet, "Timber Island: The Rosewood and Ebony Trade of Madagascar," TRAFFIC Report, November 2016.

17. Ratsimbazafy, Newton, and Ringuet, "Timber Island," 60.
18. Anna Tsing, *The Mushroom at the End of the World: On the Possibility of Life in Capitalist Ruins* (Princeton, NJ: Princeton University Press, 2015), ix.
19. This is not to say that my Western imaginary of China has entirely disappeared. My ideas of China will, of course, always reflect a Western imaginary, although this book represents a sincere effort to better understand what is going on with rosewood and other endangered species in China and Madagascar outside of what is commonly reported in Western media and scholarship.
20. Raymond Williams, *The Country and the City* (Oxford: Oxford University Press, 1975); Noel Castree and Bruce Braun, *Social Nature: Theory, Practice, Politics* (Oxford: Blackwell, 2001); Bruno Latour, *Politics of Nature: How to Bring the Sciences into Democracy* (Cambridge, MA: Harvard University Press, 2004); Elizabeth Grosz, *Time Travels: Feminism, Nature, Power* (Durham, NC: Duke University Press, 2005); Donna J. Haraway, *Staying with the Trouble: Making Kin in the Chthulucene* (Durham, NC: Duke University Press, 2016); Jennifer Clary-Lemon, *Planting the Anthropocene: Rhetorics of Natureculture* (Boulder: University Press of Colorado, 2019).
21. William Cronon, *Uncommon Ground: Rethinking the Human Place in Nature* (New York: W. W. Norton, 1995), 69–70.
22. Qijun Wang, "The Main Hall: Nucleus of the Chinese Home," in *Classical and Vernacular Chinese Furniture in the Living Environment,* ed. Kai-Yin Lo (Chicago: Art Media Resources, 1998), 96.
23. Li Yuan, "Everything Old Is New Again: Classical Chinese Furniture," *China Today,* November 17, 2011, 39.
24. Puay-peng Ho, "The Seventeenth Century House: The Dialectic of the Living Environment," in *Classical and Vernacular Chinese Furniture in the Living Environment,* ed. Kay-Yin Lo (Chicago: Art Media Resources, 1998), 57.
25. Timothy Brook, *The Confusions of Pleasure: Commerce and Culture in Ming China* (Berkeley: University of California Press, 1999), 119. A renewed loosening of maritime trade restrictions in the 1560s once again eased pressure on domestic reserves, triggering an increase in rosewood imports from the rest of Asia.
26. John Ang, "Further Studies of Furniture in Alternative Woods: Reflections on Aspects of Chinese Culture," in *Classical and Vernacular Chinese Furniture in the Living Environment,* ed. Kai-Yin Lo (Chicago: Art Media Resources, 1998), 67.
27. Joel Andreas, *Rise of the Red Engineers: The Cultural Revolution and the Origins of China's New Class* (Stanford, CA: Stanford University Press, 2009).
28. Glen Joffe, "Now You See It, Now You Don't—the Disappearance of Chinese Antique Furniture," *Primitive Living and Collecting* (blog), May 15, 2015, https://www.beprimitive.com/blog/now-you-see-it-now-you-dont-the-disappearance-of-chinese-antique-furniture.

29. This rather controversial quote (*zhìfù guāngróng*), which translates better into "wealth [to be interpreted broadly] is glorious," cannot actually be proved to be from Deng. Evelyn Iritani, "Great Idea but Don't Quote Him," *Los Angeles Times,* September 9, 2004.
30. Michel Chevalier and Pierre Xiao Lu, *Luxury China: Market Opportunities and Potential* (Singapore: John Wiley & Sons, 2010). China now serves as the biggest luxury market, with Chinese shoppers accounting for "nearly half of the global luxury market, providing invaluable demand to brands in every segment." Christopher Horton, "When It Comes to Luxury, China Still Leads," *New York Times,* April 5, 2016.
31. Wang Xin, "Desperately Seeking Status: Political, Social and Cultural Attributes of China's Rising Middle Class," *Modern China Studies* 20, no. 1 (2013): 1.
32. James Palmer, "The Balinghou: China's Generational Gap Has Never Yawned Wider," *AEON Essays,* March 7, 2013.
33. Environmental Investigation Agency, "Organised Chaos: The Illicit Overland Timber Trade between Myanmar and China" (London: Environmental Investigation Agency, 2015).
34. Importers I interviewed initially pointed out this connection to me, and I later confirmed it in news reports on Chinese timber markets after international trade restrictions were imposed.
35. Erik Swyngedouw was the first to coin the term, drawing on two distinct schools of thought: (1) critical geographers emphasizing the interconnectedness of society and nature (such as David Harvey's insistence on the "naturalness" of New York City and Neil Smith's "production of nature") and (2) science studies scholars emphasizing the hybridity among the techno-scientific and organic (such as Bruno Latour's "quasi-objects" and Donna Haraway's "cyborgs"). Erik Swyngedouw, "Modernity and Hybridity: Nature, Regeneracionismo, and the Production of the Spanish Waterscape, 1890–1930," in *Reading Economic Geography,* ed. Jamie Peck, Eric Sheppard, and Adam Tickell (Oxford: Blackwell, 2004), 189–204. My usage of the term also derives from these two schools of thought and follows Nancy Peluso's appreciation of using a socionatural approach "to better explore the entanglements of environment, commodities, and subjects." Nancy Lee Peluso, "What's Nature Got to Do with It? A Situated Historical Perspective on Socio-natural Commodities," *Development and Change* 43, no. 1 (2012): 79–104, 81.
36. Peluso, "What's Nature Got to Do with It?," 81.
37. Noel Castree, "Socializing Nature: Theory, Practice, and Politics," in *Social Nature: Theory, Practice, and Politics,* ed. Noel Castree and Bruce Braun (Oxford: Blackwell, 2001), 1–21.
38. William Cronon refers to biodiversity as a more "scientific" expression of our wild reverences. Cronon, *Uncommon Ground.*
39. David Takacs, *The Idea of Biodiversity: Philosophies of Paradise* (Baltimore, MD: Johns Hopkins University Press, 1996). In this book, Takacs refers to biodiversity as a scientific "weapon to be wielded" in pursuit of a normative "love" for the natural world (p. 3).
40. Cronon, *Uncommon Ground,* 82.
41. Curt Meine, Michael Soule, and Reed Noss, "A Mission-Driven Discipline: The Growth of Conservation Biology," *Conservation Biology* 20, no. 3 (2006): 631–651.
42. Unai Pascual et al., "Biodiversity and the Challenge of Pluralism," *Nature Sustainability* 4 (2021): 567–572; Paige West and Dan Brockington, "An Anthropological Perspective on Some Unexpected Consequences of Protected Areas," *Conservation Biology* 20, no. 3 (2006): 609–616.
43. Cronon, *Uncommon Ground,* 82.

44. Geoffrey Bowker, "Time, Money, and Biodiversity," in *Global Assemblages: Technology, Politics, and Ethics as Anthropological Problems,* ed. Aihwa Ong and Stephen J. Collier (Oxford: Blackwell, 2005): 107–123.
45. Ramachandra Guha, "Radical American Environmentalism and Wilderness Preservation: A Third World Critique," *Environmental Ethics* 11, no. 1 (1989): 93.
46. B. Aymoz et al., "Community Management of Natural Resources: A Case Study from Ankarafantsika National Park, Madagascar," *Ambio* 42 (2013): 767.
47. Russell A. Mittermeier et al., "Global Biodiversity Conservation: The Critical Role of Hotspots," in *Biodiversity Hotspots,* ed. Frank E. Zachos and Jan Christian Habel (Berlin: Springer-Verlag Berlin Heidelberg, 2011), 3–22.
48. In making this argument, I build on the work of Robert Weller in *Discovering Nature: Globalization and Environmental Culture in China and Taiwan* (Cambridge, UK: Cambridge University Press, 2006).
49. Ching Kwan Lee, *The Specter of Global China: Politics, Labor, and Foreign Investment in Africa* (Chicago: University of Chicago Press, 2017).
50. Lester Brown, *Who Will Feed China? Wake-Up Call for a Small Planet* (New York: W. W. Norton, 1995); Barbara Finamore, *Will China Save the Planet?* (Cambridge, UK: Polity Press, 2018); Yifei Li and Judith Shapiro, *China Goes Green: Coercive Environmentalism for a Troubled Planet* (Cambridge, UK: Polity Press, 2020).
51. See, for example, Emily Yeh, "Greening Western China: A Critical View," *Geoforum* 40, no. 5 (2009): 884–894.
52. Li and Shapiro, *China Goes Green,* 22.
53. In doing so, this book builds on the work of Ching Kwan Lee in *The Specter of Global China: Politics, Labor, and Foreign Investment in Africa* (Chicago: University of Chicago Press, 2018) and Deborah Bräutigam in *The Dragon's Gift: The Real Story of China in Africa* (Oxford: Oxford University Press, 2011), but with a more environmental focus.
54. This book focuses primarily on the debate in terms of endangered species and nature preservation versus sustainable use. Discussions of the global timber supply chain are not thoroughly covered in the book. For further discussion on this topic, see Nigel Dudley, Jean-Paul Jeanrenaud, and Francis Sullivan, *Bad Harvest? The Timber Trade and the Degradation of Global Forests* (New York: Routledge, 2014); Xiaoxue Weng et al., "The Africa-China Timber Trade: Diverse Business Models Call for Specialized Policy Responses" (Bogor, Indonesia: Center for International Forestry Research, 2014); Qian Zhang et al., "Global Timber Harvest Footprints of Nations and Virtual Timber Trade Flows," *Journal of Cleaner Production* 250 (2020): 119503; Y. Zhao, Y. M. Krott, and S. Ongolo, "The Governance of Global Forest Statistics: Case Study on Forest Products Trade between Sub-Saharan Africa and China," *International Forestry Review* 22, no. 3 (2020): 408–424.
55. Elena Tracy et al., "China's New Eurasian Ambitions: The Environmental Risks of the Silk Road Economic Belt," *Eurasian Geography and Economics* 58, no. 1 (2017): 56–88; William Laurance and Irene Burgués Arrea, "Roads to Riches or Ruin?," *Science* 358, no. 6362 (2017): 442–444.
56. Judith Shapiro, *Mao's War against Nature: Politics and the Environment in Revolutionary China* (Cambridge, UK: Cambridge University Press, 2001).
57. Poplar is simultaneously criticized by Western conservationists for being a fast-growing cash crop and lauded by Chinese developers for having brought economic benefits while rehabilitating landscapes in Western China.
58. Pádraig Carmody, *The New Scramble for Africa* (Cambridge, UK: Polity Press, 2017).

1. cultural bloom, cultural boom

1. "Huanghuali Furniture Made a New World Record of 70 Million," http://www.idsatool.com/info-huanghuali-furniture-made-a-new-1068.html.
2. United Nations Office on Drugs and Crime, *World Wildlife Crime Report* (New York: United Nations, 2016 and 2020).
3. Thorstein Veblen, *The Theory of the Leisure Class: An Economic Study in the Evolution of Institutions* (New York: Macmillan, 1899).
4. Jared D. Margulies, Rebecca W. Y. Wong, and Rosaleen Duffy, "The Imaginary 'Asian Super Consumer': A Critique of Demand Reduction Campaigns for the Illegal Wildlife Trade," *Geoforum* 107 (December 2019): 216–219.
5. Eileen Guo, "Chinese Consumers' Crazy Rich Demand for Rosewood Propels Drive toward Its Extinction," *South China Morning Post*, September 18, 2018.
6. Por Peal, "Dead Wood," *Economist*, December 17, 2011.
7. Yu Xiao, "China's Ivory Market after the Ivory Trade Ban in 2018," TRAFFIC Briefing (Washington, DC: World Wildlife Fund, 2018).
8. G. Burgess et al., "Reducing Demand for Illegal Wildlife Products: Research Analysis on Strategies to Change Illegal Wildlife Product Consumer Behaviour" (Cambridge, UK: TRAFFIC, University of Oxford, Imperial College Business School, and World Wildlife Fund, September 2018).
9. Michael Scott Smith, "Framing Rhino Horn Demand Reduction in Vietnam: Dismissing Medical Use as Voodoo," *Pacific Journalism Review* 24, no. 2 (2018): 241–256. See also Hubert Cheung et al., "Understanding Traditional Chinese Medicine to Strengthen Conservation Outcomes," *People and Nature* 3, no. 1 (February 2021): 115–128.
10. Hoai Nam Dang Vu and Martin Reinhardt Nielsen, "Evidence or Delusion: A Critique of Contemporary Rhino Horn Demand Reduction Strategies," *Human Dimensions of Wildlife* 26, no. 4 (2021): 390–400
11. Cheung et al., "Understanding Traditional Chinese Medicine."
12. "Wet markets" is a term used to refer to a market selling live or recently slaughtered animals. These may or may not contain wildlife. Wet markets, therefore, are distinct from wildlife markets.
13. For further discussion, see Annah Zhu and George Zhu, "Understanding China's Wildlife Markets: Trade and Tradition in an Age of Pandemic," *World Development* 136 (2020): 105108.
14. Jianmin Wen and Shi Dazhao, "Beware of the West Attacking Traditional Chinese Medicine under the Name of Wildlife Protection," *Xinhua*, March 15, 2016.
15. Zhang Hou (1511–1593) made the first remark about the excesses of the late Ming in his *Account of the Hundred Crafts*. See Craig Clunas, "Things in Between: Splendour and Excess in Ming China," in *The Oxford Handbook of the History of Consumption*, ed. Frank Trentmann (Oxford: Oxford University Press, 2012), 49. For the second quotation, see Craig Clunas and Jessica Harrison-Hall, eds., *Ming: 50 Years That Changed China* (London: British Museum Press, 2014), 11. For the third quotation, see Craig Clunas, *Superfluous Things: Material Culture and Social Status in Early Modern China* (Honolulu: University of Hawai'i Press, 2004), 169, in which Clunas refers to the late Ming as "a culturally exquisite bloom which was nevertheless in some sense rotten inside."
16. Timothy Brook, *The Confusions of Pleasure: Commerce and Culture in Ming China* (Berkeley: University of California Press, 1999), 1.

17. Clunas, "Things in Between," 49.
18. These controls were codified primarily in the Ming Code, which provided regulations for personal possessions and appearance according to class, as well as regulations for movement and migration within the dynastic boundaries.
19. Curtis Evarts, "Traditional Chinese Furniture: Sumptuous Palaces to Courtyard Dwellings," in *Classical and Vernacular Chinese Furniture in the Living Environment,* ed. Kai-Yin Lo (Chicago: Art Media Resources, 1998), 33.
20. Demonstrating this, Fan Chao (born 1540) wrote, "When I was young, I saw not a single piece of hardwood furniture. . . . The people only had square tables of gingko wood or gilded lacquer." Noting the proliferation of hardwood furniture over time, he continued, "This furniture was extremely expensive and fine, costing tens of thousands of cash, an extravagant custom." Clunas, *Superfluous Things,* 155.
21. Ibid., 171.
22. Ibid.
23. Jonathan Hay notes, "What made taste a compelling issue in the late Ming was the realisation that social status was not after all, immutable" (339). Jonathan Hay, *Shitao: Painting and Modernity in Early Qing China* (Cambridge: Cambridge University Press, 2001).
24. This opposition was used by many cultural authorities during the late Ming (including Wen Zhenheng, Li Yu, Gao Lian, and Ji Cheng). Puay-peng Ho, "The Seventeenth Century House: The Dialectic of the Living Environment," in *Classical and Vernacular Chinese Furniture in the Living Environment,* ed. Kai-Yin Lo (Chicago: Art Media Resources, 1998).
25. Most seminal among them is Wen Zhenheng's *Treatise on Superfluous Things* (c. 1620–1627), but see also Gao Lian's *Eight Discourses on the Art of Living* (1591) and Tu Long's *Desultory Remarks on Furnishing* (1606), as well as the following multiauthored collections: *Necessities for the Householder* (c. 1600–1640), *Encyclopedia for Daily Use* (c. 1580–1640), and *Complete and Categorized Essentials for the Householder* (n.d.). Craig Clunas notes, "What is distinctive about China, however, is not that there was an uneven distribution of knowledge about how to deploy consumption to achieve social ends. . . . The Chinese distinction lies in the very early reduction of this type of [elite consumer] knowledge to a commodity published in a book and hence available in a marketplace to any player wanting to enter the search for ways of transforming economic power into cultural power." Clunas, *Superfluous Things,* 12–13.
26. Ho, "Seventeenth Century House."
27. This was at a time when foot-binding was widely practiced. This painful tradition distorted women's feet to appear small and dainty from the outside, but they were mutilated and often infected beneath their bandages. Good scholarly men, on the other hand, required footrests.
28. Indeed, modestly carved hardwood constituted the "perfected balance between elegance of decoration and solidity of structure" that defined the aesthetics of the age. Evarts, "Traditional Chinese Furniture."
29. Sophia Kappus, "China's Ivory: An Approach to the Conflict between Tradition and Ethical Responsibility" (master's thesis, Leiden University, 2018).
30. Michel Bonnin, *The Lost Generation: The Rustication of China's Educated Youth (1968–1980),* trans. Krystyna Horko (Hong Kong: Chinese University Press, 2013).
31. Andrew G. Walder, *China under Mao: A Revolution Derailed* (Cambridge, MA: Harvard University Press, 2015), 50.

32. Walder, *China under Mao*, 55. Although the initial land reforms generated widespread peasant support, the subsequent collectivization reforms registered local resistance, hardship, and hunger.
33. "This revolution," Joel Andreas writes, "was presented as a natural continuation of the Communist class leveling program, extending it from the economic to the cultural field. Having eliminated the economic dominance of the old elite classes, the CCP could now turn its attention to their cultural dominance." Joel Andreas, *Rise of the Red Engineers: The Cultural Revolution and the Origins of China's New Class* (Stanford, CA: Stanford University Press, 2009), 49.
34. Roger Schwendeman, "Burn, Loot and Pillage! Destruction of Antiques during China's Cultural Revolution," *ACF China* (blog), February 10, 2013, http://www.antique-chinese-furniture.com/blog/2013/02/10/burn-loot-and-pillage-destruction-of-antiques-during-chinas-cultural-revolution/. At the time I was interviewing, in 2014 and 2015, the few families that did purchase (very cheap) rosewood furniture during the Cultural Revolution found the price of their possessions increase by many orders of magnitude after 2005.
35. Weiwei Zhang, "No Cultural Revolution? Continuity and Change in Consumption Patterns in Contemporary China," *Journal of Consumer Culture* 17, no. 3 (October 18, 2015): 639–658.
36. Joel Andreas also makes this point in noting that cultural traditions "attracted suspicion as markers of the 'feudal' and 'bourgeois' culture." Andreas, *Rise of the Red Engineers*, 46.
37. Juan Du, *The Shenzhen Experiment: The Story of China's Instant City* (Cambridge, MA: Harvard University Press, 2020).
38. Ronald Coase and Ning Wang, *How China Became Capitalist* (Basingstoke, UK: Palgrave Macmillan, 2012), 66. Concerning the lack of ideological sway, Coase and Wang write, "Without a clear vision of where Chinese should be heading, post-Mao Chinese leaders opened their mind."
39. As noted in the introduction, these individuals are referred to as *bàofā hù* ("explosively rich"), a term that carries the "denigrating connotation of being uncultured and without cultural worth." Wang Xin, "Desperately Seeking Status: Political, Social and Cultural Attributes of China's Rising Middle Class," *Modern China Studies* 20, no. 1 (2013): 36.
40. World Bank Data Catalog, "World Development Indicators: GDP per Capita (Current US$)," accessed February 2021, https://datacatalog.worldbank.org/dataset/world-development-indicators.
41. Michael F. Martin and Wayne M. Morrison, "China's 'Hot Money' Problems," *CRS Report for Congress*, July 21, 2008, https://fas.org/sgp/crs/row/RS22921.pdf.
42. Barry Eichengreen, "Hot Money: The True Meaning of China's Capital Controls," Centre for International Governance Innovation, December 13, 2016, https://www.cigionline.org/articles/hot-money-true-meaning-chinas-capital-controls.
43. Sandy Li and Pearl Liu, "Shanghai Changes Rules to Stop Couples from Faking Divorces as They Vie for Less Up-Front Money to Buy Residential Property," *South China Morning Post*, January 22, 2021.
44. Yan Yiqi, "Rosewood Furniture Increases in Value over Time," *China Daily*, January 12, 2015.
45. Li Yuan, "Everything Old Is New Again: Classical Chinese Furniture," *China Today*, November 17, 2011.
46. "Rosy Times Ahead for Rosewood Furniture," *China Daily*, January 17, 2011.

47. "Rosewood Furniture in Price Heat," *Report from China International Rosewood Classical Furniture Exhibition,* September 30, 2010.
48. Yuan, "Everything Old Is New Again," 41.
49. Huang Wenbin and Sun Xiufang, "Tropical Hardwood Flows in China: Case Studies of Rosewood and Okoumé" (Washington, DC: Forest Trends Association, 2013); Environmental Investigation Agency, *The Hongmu Challenge: A Briefing for the 66th Meeting of the CITES Standing Committee* (London: Environmental Investigation Agency, January 11, 2016).
50. Naomi Basik Treanor, "China's *Hongmu* Consumption Boom: Analysis of the Chinese Rosewood Trade and Links to Illegal Activity in Tropical Forested Countries," Forest Trends Report Series (Washington, DC: Forest Trends Association, December 2015).
51. Five interviewees working at the Shanghai Furen Timber Market commented on this trend (December 2015). The same exact phrase (*yǒu jià wú shì*) was used by traders to describe the rhino horn market, which suffered from a similar slowdown in market turnover but nonetheless retains a high price. Apparently, "collectors, investors, and speculators are holding onto their collections, refusing to sell at a low price and waiting for the policy to change." Yufang Gao et al., "Rhino Horn Trade in China: An Analysis of the Art and Antiques Market," *Biological Conservation* 201 (2016): 343–347, 346.
52. Amy Nip, "'Chicken Cup' Sets HK$281m World Record for Chinese Porcelain at Hong Kong Auction," *South China Morning Post,* April 8, 2014.
53. "Chinese Ru-ware Bowl Sets $38m Auction Record in Hong Kong," BBC News, October 3, 2017.
54. "Chinese Vase Fetches Record $69 Million in UK Auction," Reuters, November 12, 2010. The article also notes that "the sale highlights the intense and growing competition among wealthy Chinese buyers for rare pieces of their heritage, and anything associated with imperial China appears to be particularly attractive."
55. Alex W. Palmer, "The Great Chinese Art Heist," *GQ,* August 16, 2018.
56. Larissa Buchholz, "What Is a Global Field? Theorizing Fields beyond the Nation-State," *Sociological Review* 64, no. 2 (March 2016): 31–60.
57. "Contemporary Art Migrates to the East," Kooness, February 26, 2019.
58. "Christie's Contemporary HK Sale to Benefit from Global Auction Trends," *Jing Daily,* November 20, 2012.
59. "A 'Huanghuali' Corner-Leg Table Sold for £3m, 50 Times Its Estimate," *The Value,* November 8, 2018.
60. Eric Hobsbawm and Terence Ranger, eds., *The Invention of Tradition* (Cambridge, UK: Cambridge University Press, 2012).

2. hot money capitalism

1. Susan Strange, *Casino Capitalism* (Manchester, UK: Manchester University Press, 1998).
2. Karl Marx, *Capital,* vol. 5, *A Critique of Political Economy* (London: Penguin Classics, 1992).
3. Michael Taussig, *The Devil and Commodity Fetishism in South America* (Chapel Hill: University of North Carolina Press, 1980), 10.
4. Aihwa Ong, *Spirits of Resistance and Capitalist Discipline: Factory Women in Malaysia,* 2nd ed. (Albany: State University of New York Press, 2010); Peter Geschiere, *The Modernity of Witchcraft: Politics and the Occult in Postcolonial Africa* (Charlottesville: University of Virginia Press, 1997).

5. Jean Comaroff and John L. Comaroff, eds., *Modernity and Its Malcontents: Ritual and Power in Postcolonial Africa* (Chicago: University of Chicago Press, 1993), xiv.
6. Geschiere, *Modernity of Witchcraft.*
7. Taussig, *Devil and Commodity Fetishism.*
8. June Nash, *We Eat the Mines and the Mines Eat Us: Dependency and Exploitation in Bolivian Tin Mines* (New York: Columbia University Press, 1979), 15.
9. Ibid.
10. Sidney W. Mintz, *Sweetness and Power: The Place of Sugar in Modern History* (New York: Viking Penguin, 1985); Sarah Lyon and Mark Moberg, eds., *Fair Trade and Social Justice: Global Ethnographies* (New York: New York University Press, 2010); William Roseberry, Lowell Gudmundson, and Mario Samper Kutschbach, eds., *Coffee, Society, and Power in Latin America* (Baltimore: Johns Hopkins University Press, 1995). Steve Striffler and Mark Moberg, eds., *Banana Wars: Power, Production, and History in the Americas* (Durham, NC: Duke University Press, 2003).
11. James Ferguson, *Global Shadows: Africa in the Neoliberal World Order* (Durham, NC: Duke University Press, 2006).
12. Anna Lowenhaupt Tsing, *The Mushroom at the End of the World: On the Possibility of Life in Capitalist Ruins* (Princeton, NJ: Princeton University Press, 2015).
13. David Cleary, *Anatomy of the Amazon Gold Rush* (Iowa City: University of Iowa Press, 1990); Filip De Boeck, "Domesticating Diamonds and Dollars: Identity, Expenditure and Sharing in Southwestern Zaire (1984–1997)," in *Globalization and Identity: Dialectics of Flow and Closure,* ed. Birgit Meyer and Peter Geschiere (Oxford: Blackwell, 1999), 177–210; Stephen Jackson, "Fortunes of War: The Coltan Trade in the Kivus," Overseas Development Institute Working Paper, background research for HPG Report 13, February 2003.
14. Andrew Walsh, "'Hot Money' and Daring Consumption in a Northern Malagasy Sapphire-Mining Town," *American Ethnologist* 30, no. 2 (2003): 290–305; Genese Marie Sodikoff, *Forest and Labor in Madagascar: From Colonial Concession to Global Biosphere* (Bloomington: Indiana University Press, 2012); Jennifer Cole, "Fresh Contact in Tamatave, Madagascar: Sex, Money, and Intergenerational Transformation," *American Ethnologist* 31, no. 4 (November 2004): 573–588; Jennifer Cole, "The Jaombilo of Tamatave (Madagascar), 1992–2004: Reflections on Youth and Globalization," *Journal of Social History* 38, no. 4 (Summer 2005): 891–914. A similar practice of rash spending has also been documented in the Vezo fishing communities of southwestern Madagascar, where fishers may spend their entire earnings from a good catch with little planning for the future. See Rita Astuti, "At the Center of the Market: A Vezo Woman," in *Lilies of the Field: Marginal People Who Live for the Moment,* ed. Sophie Day, Evthymios Papataxiarchis, and Michael Stewart (Boulder, CO: Westview Press, 1999), 83–95.
15. For the former, see Oscar Lewis, "The Culture of Poverty," *Scientific American* 215, no. 4 (October 1966): 19–25. For the latter, see Claude Meillassoux, "Ostentation, Destruction, Reproduction," *Economies et Sociétés* 2, no. 4 (1968): 760–772, and James Woodburn, "Egalitarian Societies," in *Limited Wants, Unlimited Means: A Reader on Hunter-Gatherer Economics and the Environment,* ed. John Gowdy (Washington, DC: Island Press, 1998), 87–110.
16. Sophie Day, Evthymios Papataxiarchis, and Michael Stewart, eds., *Lilies of the Field: Marginal People Who Live for the Moment* (Boulder, CO: Westview Press, 1999).
17. Sasha Newell, *The Modernity Bluff: Crime, Consumption, and Citizenship in Côte d'Ivoire* (Chicago: University of Chicago Press, 2012), 99; Walsh, "'Hot Money' and Daring Consumption," 299.

18. Tsing, *Mushroom at the End of the World.*
19. Sodikoff, *Forest and Labor in Madagascar;* Newell, *Modernity Bluff.*
20. Although some of the best paved roads in the country connect the region's coastal cities, the roads leading in and out of the region take days to travel by bush taxi and are often quite dangerous.
21. This, however, may change. China recently funded a long-overdue project to pave the road across northern Madagascar connecting Vohemar to Ambilobe (see Figure 2.1).
22. Sarah Osterhoudt, *Vanilla Landscapes: Meaning, Memory and the Cultivation of Place in Madagascar* (New York: New York Botanical Garden Press, 2017), 47.
23. Lynn Pan, ed., *The Encyclopedia of the Chinese Overseas* (Cambridge, MA: Harvard University Press, 1999).
24. Sodikoff, *Forest and Labor in Madagascar.*
25. Patricia Rain, *Vanilla: The Cultural History of the World's Most Popular Flavor and Fragrance* (London: Penguin, 2004), 308.
26. World Bank, "Governance and Development Effectiveness Review: A Political Economy Analysis of Governance in Madagascar," World Bank Report No. 54277-MG (2010), xiii.
27. "La 'bolabolacratie,'" *Madagascar Tribune,* March 1, 2012.
28. "A Madagascar, le prix de la vanille flambe, sa qualité se dégrade," AFP, May 8, 2016.
29. In interviews conducted during the summers of 2014 and 2015, respondents often commented that when the price of rosewood was high, the price of vanilla was low, and vice versa.
30. Fredric Jameson, *Postmodernism, or, The Cultural Logic of Late Capitalism* (Durham, NC: Duke University Press, 1991); Ernest Mandel, *Late Capitalism* (London: Verso Classics, 1999 [1975]).
31. Fredric Jameson, "Culture and Finance Capital," *Critical Inquiry* 24, no. 1 (Autumn 1997): 260.
32. "Far from being inconsistent with Marx's great 19th Century analysis," Jameson writes of late capitalism, "[it] constitutes on the contrary the purest form of capital yet to have emerged." Jameson, *Postmodernism,* 78.
33. Ibid., 251.
34. Margaret L. Brown, "Madagascar's Cyclone Vulnerability and the Global Vanilla Economy," in *The Political Economy of Hazards and Disasters,* ed. Eric C. Jones and Arthur D. Murphy (Lanham, MD: AltaMira Press, 2009), 253–254.
35. Rain, *Vanilla,* 229.
36. James Altucher, "Supply, Demand, and Edible Orchids," *Financial Times,* September 20, 2005, 12.
37. Monte Reel, "Vanillanomics," *Bloomberg Businessweek,* December 17, 2019.
38. Ibid.
39. Tim Ecott, *Vanilla: Travels in Search of the Ice Cream Orchid* (New York: Grove Press, 2004).
40. Rebecca Burn-Callander, "Vanilla Shortage Hits Ice Cream Makers," *Management Today,* April 3, 2012.
41. Related to, but not to be confused with, *cultures d'exportation* (export crops).
42. Jonathan Parry and Maurice Bloch, eds., *Money and the Morality of Exchange* (Cambridge, UK: Cambridge University Press, 1989); Étienne Flaucourt, *Histoire de la Grand Île de Madagascar,* 1661.
43. Jennifer Cole, *Forget Colonialism? Sacrifice and the Art of Memory in Madagascar* (Berkeley: University of California Press), 201.
44. Ecott, *Vanilla.*

45. Cole, *Forget Colonialism?*, 196.
46. Mintz, *Sweetness and Power*, 109.
47. Barry Bearak, "Shaky Rule in Madagascar Threatens Trees," *New York Times*, May 24, 2010.
48. Ecott, *Vanilla*.
49. Maurice Bloch, *Placing the Dead: Tombs, Ancestral Villages, and Kinship Organization in Madagascar* (London: Seminar Press, 1971); David Graeber, "Dancing with Corpses Reconsidered: An Interpretation of 'Famadihana' (in Arivonimamo, Madagascar)," *American Ethnologist* 22, no. 2 (May 1995): 258–278.
50. Walsh, "'Hot Money' and Daring Consumption," 299.
51. Day, Papataxiarchis, and Stewart, *Lilies of the Field*, 13.

3. taking back the forest

1. Referred to as *lundi noir* (French) or *alatsinainy mainty* (Malagasy).
2. "Un an de désastres," *Madagascar Tribune*, January 26, 2010.
3. Derek Shuurman and Porter P. Lowry II, "The Madagascar Rosewood Massacre," *Madagascar Conservation & Development* 4, no. 2 (December 2009): 98–102; Hery Randriamalala and Z. Liu, "Bois de rose de Madagascar: Entre démocratie et protection de la nature," *Madagascar Conservation & Development* 5, no. 1 (June 2010): 11–22; Global Witness and Environmental Investigation Agency, "Investigation into the Illegal Felling, Transport, and Export of Precious Wood in SAVA Region Madagascar" (London: Global Witness and Environmental Investigation Agency, August 2009); World Bank, "Governance and Development Effectiveness Review: A Political Economy Analysis of Governance in Madagascar," World Bank Report No. 54277-MG (2010); Hery Randriamalala, "Rosewood Chronicles," *Illegal Logging Portal*, Forest Legality Initiative, August 22, 2011; "Un an de désastres"; Environmental Investigation Agency, "The Ongoing Illegal Logging Crisis in Madagascar: An EIA Briefing for CITES SC65," July 18, 2014; "Anthelme Ramparany: Une ministre de trafic du bois de rose?," *TanaNews*, April 22, 2014.
4. "Un an de désastres"; Randriamalala, "Rosewood Chronicles"; Global Witness and Environmental Investigation Agency, "Investigation into the Illegal Felling."
5. Exactly how much cached rosewood the region is sitting on is wildly speculative. Estimates range from $600 million to $5 billion (the latter cited in an open letter signed by forty organizations, including a number of major nongovernmental organizations). The figure depends largely on the value of rosewood in China, which vacillates sharply according to market speculation, as discussed in Chapters 1 and 5.
6. This oppression began first with the Merina monarchy, ruling from the Central Highlands, and continued with the French colonial regime and postcolonial power structures receiving mandates from abroad (see Chapter 4).
7. "Un an de désastres."
8. Indeed, the only prominent arrests concerning rosewood are of the few locals who have spoken out about the trade and facilitated its international exposure. Rowan Moore Gerety, "Activist Arrested While Illegal Loggers Chop Away at Madagascar's Forests," Mongabay, September 15, 2015, https://news.mongabay.com/2015/09/activist-arrested-while-illegal-loggers-chop-away-at-madagascars-forests/.
9. Maurice Bloch and Jonathan Parry, eds., *Money and the Morality of Exchange* (Cambridge, UK: Cambridge University Press, 1989), 184.

10. Ibid.
11. Ibid., 18.
12. Nancy Lee Peluso and Peter Vandergeest, "Genealogies of the Political Forest and Customary Rights in Indonesia, Malaysia, and Thailand," *Journal of Asian Studies* 60, no. 3 (August 2001): 761–812.
13. Eva Keller, "The Banana Plant and the Moon: Conservation and the Malagasy Ethos of Life in Masoala, Madagascar," *American Ethnologist* 35, no. 4 (November 2008): 650–664.
14. William Reno, "Clandestine Economies, Violence and States in Africa," *Journal of International Affairs* 53, no. 2 (Spring 2000): 433–459, 434.
15. Jean-François Bayart, *The State in Africa: The Politics of the Belly* (Malden, MA: Polity, 2009).
16. James Ferguson, *Global Shadows: Africa in the Neoliberal World Order* (Durham, NC: Duke University Press, 2006).
17. See, for example, Sarah Milne, "Cambodia's Unofficial Regime of Extraction: Illicit Logging in the Shadow of Transnational Governance and Investment," *Critical Asian Studies* 47, no. 2 (2015): 200–228.
18. Sarinda Singh, "Borderland Practices and Narratives: Illegal Cross-Border Logging in Northeastern Cambodia," *Ethnography* 15, no. 2 (2014): 135–159; Phuc Xuan To, Sango Mahanty, and Wolfram Dressler, "Social Networks of Corruption in the Vietnamese and Lao Cross-Border Timber Trade," *Anthropological Forum* 24, no. 2 (2014): 154–174.
19. The entire process is remarkably manual, with little machine involvement. Machines are not used primarily because they are too expensive and difficult to haul into remote areas and labor is cheap, but also because the work is illegal and performed as discreetly as possible, without the noise or visibility of machinery.
20. Entry-level vanilla drying and sorting positions typically earn workers less than 2,000 ariary less than (US$1) per day. The legal minimum wage for agricultural work is the equivalent of 18 cents per hour.
21. *Veloma baba* is a name given to places associated with other dangerous acts, such as mining. See Andrew Walsh, "'Hot Money' and Daring Consumption in a Northern Malagasy Sapphire-Mining Town," *American Ethnologist* 30, no. 2 (2003): 290–305.
22. Approximately 90 percent of the houses were destroyed, leaving more than one hundred thousand people homeless and nearly three hundred thousand in need of immediate assistance. With waves towering at eight meters, the coastline became permanently reconfigured after the storm surge receded. Residents in the countryside recalled a long night of huddling under the strongest wood houses and emerging early in the morning to find only the floor remaining. In the city, many sought refuge in the few public concrete structures available. Some of the men, fearing theft, weathered the storm in their homes, huddled in the corner as the wind tore off their roofs. Reports covering the aftermath of the storm described the city of Antalaha as "virtually erased" (International Federation of Red Cross and Red Crescent Societies, "Madagascar: Cyclones and Floods," Appeal No. 06 / 2000, Situation Report No. 1, June 14, 2000). This devastation was experienced during a time of recovery from two cyclones that hit earlier in the season.
23. Order 11832 / 2000 banned the export of unfinished woods and suspended the issuance of licenses to operate in the region, but it also permitted salvage logging. Order 12704 / 2000 banned the extraction of wood resources in protected areas and their peripheral zones.
24. Despite some attempts to curb the logging (in March 2005, the government issued a regional decree, 001 2005 REG / SAV, prohibiting traffic in rosewood and mobilizing inspection

brigades to appease World Bank loan conditions), in October 2005, facing an upcoming election period, the Malagasy government capitulated with the grievances of the rosewood exporters and authorized the export of existing precious wood stocks (Memorandum 923).

25. This decree permitted thirteen timber exporters to unconditionally export raw rosewood, ebony, and palisander until April 30, 2009. The deadline was later extended by the transitional regime that replaced Ravalomanana.
26. Randriamalala and Liu, "Bois de rose de Madagascar." The authors also estimate that as much as US$52 million in rosewood earnings has been delivered to overseas bank accounts and has yet to be repatriated to Madagascar.
27. Interministerial Decrees 38244 / 2009 and 38409 / 2009.
28. The World Bank estimates that the transition government received US$18–$40 million in imposed fines between September 2009 and March 2010 alone (amounting to approximately 5–10 percent of the government's revenue in 2009). World Bank, "Governance and Development Effectiveness Review."
29. Examples include the World Bank's US$52 million conservation grant, given the explicit condition that rosewood logging legislation be enforced, and the UNESCO World Heritage Committee's recommendation F 35 COM 7A.10, calling on the Malagasy government to take control of logging in the northeast.
30. "La 'bolabolacratie,'" *Madagascar Tribune,* March 1, 2012.
31. He discussed his involvement with the trade in an interview on a popular Malagasy program (available at https://www.youtube.com/watch?v=44RlEJAXFjo).
32. "Bois de rose: 62 conteneurs saisis au Kenya et au Sri Lanka," *La Gazette,* May 30, 2014.
33. Environmental Investigation Agency, "Ongoing Illegal Logging Crisis in Madagascar"; "Anthelme Ramparany: Une ministre de trafic du bois de rose?"
34. When I returned for fieldwork in May 2014, I spoke with dozens of loggers and traders who had just returned from the forest after the military arrival in March. They alleged that the trade continued, but only for those who controlled it from the capital.
35. The price of rosewood in the region was consistently cited by my interlocutors to be an average of 2,000 ariary per kilogram from 2014 to 2015, as opposed to more than four times that at the peak of the trade from 2009 to 2013.
36. The president of the Courts of Justice in one northeastern city "affirmed that 'serious legal uncertainty' had resulted in several dismissals of charges brought against exporters and officials." See Global Witness and Environmental Investigation Agency, "Investigation into the Illegal Felling," 11.
37. All rosewood from Madagascar was listed under the Convention on International Trade in Endangered Species of Wild Fauna and Flora (CITES) Appendix II, thus prohibiting trade without an export permit issued by the CITES Secretariat (none of which have been issued).
38. "Le périple illicite des bois de rose de Singapour," *L'Express de Madagascar,* February 26, 2016; "Dossier Singapour—Ramparany tait les questions sensibles," *L'Express de Madagascar,* March 11, 2016. The new minister of environment later denied the legality of the shipment, contradicting the former minister's confirmation.
39. The acquittal was a reversal of a conviction issued in 2017, sentencing the responsible party to jail time and imposing US$1 million in fines. Edward Carver, "Singapore Acquits Trader in World's Biggest Rosewood Bust, Worth $50m," Mongabay, April 19, 2019, https://news.mongabay.com/2019/04/singapore-acquits-trader-in-worlds-biggest-rosewood-bust-worth-50m/.

40. "Anthelme Ramparany: Une ministre de trafic du bois de rose?"; "Trafic de bois rose—Le verrouillage aux ports amplifie la sortie clandestine des rondins," *L'Express de Madagascar,* April 13, 2016. The convictions that *have* surfaced over the years are primarily of lower-level traders and activists speaking out against the trade.
41. "Les députés ignorent leur véritable role," *Madagascar Tribune,* October 28, 2014.
42. "La 'bolabolacratie.'"

4. worst-case conservation

1. World Environment Day is a United Nations–sponsored environmental awareness campaign, held on June 5 of every year since 1974.
2. In one instance that was recounted to me by one of my interlocutors, a gendarme was rumored to have spent all his money earned from fines during a single night at the bar. Finding himself with no money, the gendarme drunkenly returned to one of the rosewood bosses he had fined earlier, demanding further payment. Trying to calm the gendarme, the boss told him he could have his money but after he had sobered. The gendarme was said to have then killed the man while twenty of his workers watched in shock. In a separate incident, another gendarme allegedly broke into a rosewood boss's house with the help of the boss's guard and stole 2 billion ariary (about US$500,000) in cash. One of the workers staying in the house at the time was shot and killed in the process. In both cases, residents claim that the penalty for the gendarmes was a fine and relocation.
3. Arun Agrawal, *Environmentality: Technologies of Government and the Making of Subjects* (Durham, NC: Duke University Press, 2005), 16.
4. Douglas William Hume, "Swidden Agriculture and Conservation in Eastern Madagascar: Stakeholder Perspectives and Cultural Belief Systems," *Conservation & Society* 4, no. 2 (April–June 2006): 287–303.
5. "Merina" refers to the people ruling, while "Imerina" refers to the kingdom itself. Merina king Andrianampoinimerina is typically credited with beginning the political unification of the country in the late eighteenth century, as discussed in Chapter 3. King Andrianampoinimerina was succeeded by his son, Radama I, who extended Merina authority to cover nearly the entire island. Both kings spent much of their rule monopolizing the island's export industries and instituting forced-labor campaigns to assist with export operations. Precious hardwoods, including rosewood, palisander, and ebony, featured among the primary exports. Massive forced-labor campaigns were deployed to transport the logs from the interior of the forest to the ports. Imperial authorities granted foreign traders the rights to exploit these timber resources. Gwyn Campbell, *An Economic History of Imperial Madagascar, 1750–1895: The Rise and Fall of an Island Empire* (Cambridge, UK: Cambridge University Press, 2005), 129; Sandra J. T. M. Evers, Gwyn Campbell, and Michael Lambek, eds., *Contest for Land in Madagascar: Environment, Ancestors and Development* (Leiden: Brill, 2013).
6. Catherine Corson, *Corridors of Power: The Politics of Environmental Aid to Madagascar* (New Haven, CT: Yale University Press, 2016), 36.
7. This shift was perhaps not as profound in northern Madagascar as in other parts of the island. Precisely because northerners were the unwilling subjects of Merina rule, they tended to cooperate with the French against the Merina. As the French colonial regime strengthened, however, the French became increasingly viewed as the new oppressors.
8. Corson, *Corridors of Power,* 37.

9. Forest concessions reached 101,630 hectares by 1902 and 600,000 hectares by 1921. Alain Bertrand, "The Spread of the Merina People in Madagascar and Natural Forest and Eucalyptus Stand Dynamics," in *Beyond Tropical Deforestation: From Tropical Deforestation to Forest Cover Dynamics and Forest Development,* ed. Didier Babin (Montpellier: CIRAD, 2004), 151–156; Sherry H. Olson, "The Robe of the Ancestors: Forests in the History of Madagascar," *Journal of Forest History* 28, no. 4 (October 1984): 174–186.
10. Solofo Randrianja and Stephen Ellis, *Madagascar: A Short History* (Chicago: University of Chicago Press, 2009).
11. Eva Keller, *Beyond the Lens of Conservation: Malagasy and Swiss Imaginations of One Another* (New York: Berghahn Books, 2015).
12. Olson, "Robe of the Ancestors," 182.
13. Genese Marie Sodikoff, *Forest and Labor in Madagascar: From Colonial Concession to Global Biosphere* (Bloomington: Indiana University Press, 2012), 13.
14. Corson, *Corridors of Power,* 47.
15. Lucy Jarosz, "Defining Deforestation in Madagascar," in *Liberation Ecologies: Environment, Development, Social Movements,* ed. Richard Peet and Michael Watts (London: Routledge, 1996), 148–164.
16. Catherine Corson, "Territorialization, Enclosure and Neoliberalism: Non-state Influence in Struggles over Madagascar's Forests," *Journal of Peasant Studies* 38, no. 4 (2011): 703–726, 709.
17. Corson, *Corridors of Power,* 14, quoting Henri Humbert, "Parcs Nationaux et Reserves Naturelles en Afrique et a Madagascar," *Bulletin de l'Association Française pour l'Avancement des Sciences* (1933), 212.
18. Sodikoff, *Forest and Labor in Madagascar,* 93.
19. Jarosz, "Defining Deforestation in Madagascar," 155, quoting D. Whittlesey, "Shifting Cultivation," *Economic Geography* 13 (1937): 35–52.
20. Christian A. Kull, "Madagascar Aflame: Landscape Burning as Peasant Protest, Resistance, or a Resource Management Tool?," *Political Geography* 21, no. 7 (September 2002): 927–953.
21. Corson, *Corridors of Power;* Randrianja and Ellis, *Madagascar;* Keller, *Beyond the Lens of Conservation.*
22. For further reading on the history of postcolonial Madagascar, see Jennifer Cole, *Forget Colonialism? Sacrifice and the Art of Memory in Madagascar* (Berkeley: University of California Press, 2001); Daniela B. Raik, "Forest Management in Madagascar: An Historical Overview," *Madagascar Conservation & Development* 2, no. 1 (2007); Lesley A. Sharp, *The Sacrificed Generation: Youth, History, and the Colonized Mind in Madagascar* (Berkeley: University of California Press, 2002).
23. Sodikoff, *Forest and Labor in Madagascar;* Randrianja and Ellis, *Madagascar;* Corson, *Corridors of Power,* 52–53.
24. Randrianja and Ellis, *Madagascar,* 195.
25. Ibid.
26. Thomas F. Allnutt et al., "Mapping Recent Deforestation and Forest Disturbance in Northeastern Madagascar," *Tropical Conservation Science* 6, no. 1 (2013): 1–15, 2.
27. Corson, "Territorialization, Enclosure and Neoliberalism," 715.
28. Jim Igoe and Dan Brockington, "Neoliberal Conservation: A Brief Introduction," in *The Environment in Anthropology: A Reader in Ecology, Culture, and Sustainable Living,* ed. Nora Haenn, Richard R. Wilk, and Allison Harnish (New York: New York University Press, 2016), 328.
29. Corson, "Territorialization, Enclosure and Neoliberalism."

30. Mananara Nord National Park is the only other park in Madagascar that has experienced similar levels of intensive rosewood logging. This park is farther south and outside of the study area (but can be found on the map provided in Figure 2.1).
31. Sodikoff, *Forest and Labor in Madagascar;* Keller, *Beyond the Lens of Conservation.*
32. This contract is part of USAID Madagascar's larger Conservation and Communities project. "USAID Awards Tetra Tech US$22 Million Contract for Biodiversity Conservation and Community Development in Madagascar," Business Wire, June 21, 2018.
33. This number is contested and depends on whether or not the cities of Antalaha and Maroantsetra are included. Keller, *Beyond the Lens of Conservation,* 150, n. 2.
34. Claire Kremen et al., "Designing the Masoala National Park in Madagascar Based on Biological and Socioeconomic Data," *Conservation Biology* 13, no. 5 (October 1999): 1055–1068, 1065.
35. Households were estimated to require five hectares for forest product collection and were afforded no additional land to practice swidden agriculture. From the GPS units collected of current village territories and the estimation of future fuelwood use, the borders of the park were mapped. A total of forty-seven households were relocated to permanent villages to accommodate the boundary. These permanent villages are now labeled as "zones of controlled occupation" within the park and are strictly regulated. Kremen et al., "Designing the Masoala National Park," 1061–1065.
36. Half of these fees were allocated to local management committees, called COGES (*Comité de Gestion*), to fund development projects of their choice. For example, in 1999, COGES funding was used to improve roads, build school tables, construct wells and public toilets, and rehabilitate buildings. This was all performed on a minimal budget of US$700. In 2001, there was only about US$500 to fund development projects. Compare these funds with the park's annual operating costs of US$300,000–$400,000. Also compare these funds with the loss of productive land experienced by villagers living at the park borders. Moreover, these funds are not typically used in peripheral villages, which suffer the most from loss of productive land, but rather in areas of high ecotourism and comparatively higher wealth, such as the city of Maroantsetra. Alison Ormsby and Kathryn Mannle, "Ecotourism Benefits and the Role of Local Guides at Masoala National Park, Madagascar," *Journal of Sustainable Tourism* 14, no. 3 (2006): 271–287; Keller, *Beyond the Lens of Conservation,* 4.
37. "Both biological and socioeconomic data therefore pointed to the same solution," the park designers noted, "to include the large and environmentally heterogeneous core zone to protect the unique biodiversity of the area and to protect the forests of the peripheral zone through community-based economic incentives rather than legal mechanisms." Kremen et al., "Designing the Masoala National Park," 1064.
38. Keller, *Beyond the Lens of Conservation,* 123.
39. Alison Ormsby and Beth A. Kaplin, "A Framework for Understanding Community Resident Perceptions of Masoala National Park, Madagascar," *Environmental Conservation* 32, no. 2 (June 2005): 156–164.
40. In some cases, antagonism toward the park has come from the further extension of park boundaries in the directions of the villages. Eva Keller notes that in one village just south of the park, the boundary changed without local consultation (*Beyond the Lens of Conservation,* 126–130). When questioned, Madagascar National Parks claimed that the first boundary was simply a proposition, which was later rejected and replaced with a boundary that better matched the official design. The boundary relocation suddenly placed subsistence land within park boundaries. To address the situation, the park director drafted a handwritten, not

legally valid "convention" that agreed to let residents continue to plant on those areas they had previously cultivated despite the fact that they were now technically inside the park. No compensation was provided, except in four cases. In three of these four cases, the compensation villagers recalled receiving was less than that officially recorded. Arrests were made, however, when new land within the recently delimited area was cleared. Sentences ranged from a month to five years, although they often could be reduced through "payments."

41. Zuzana Burivalova et al., "Relevance of Global Forest Change Data Set to Local Conservation: Case Study of Forest Degradation in Masoala National Park, Madagascar," *Biotropica* 47, no. 2 (March 2015): 267–274.
42. Cynthia Ratsimbazafy, D. Newton, and S. Ringuet, "Timber Island: The Rosewood and Ebony Trade of Madagascar," TRAFFIC Report, November 2016, 58.
43. Burivalova et al., "Relevance of Global Forest Change Data Set to Local Conservation."
44. Ratsimbazafy, Newton, and Ringuet, "Timber Island," 60.
45. Referred to as *communauté de base* (COBA) or *vondron'olona ifotony* (VOI).
46. Daniel Austin and Hilary Bradt, *Madagascar* (Chesham, UK: Bradt Travel Guides, 2017), 329.
47. Hery Randriamalala, "Rosewood Chronicles," *Illegal Logging Portal,* Forest Legality Initiative, August 22, 2011.
48. "Destruction Worsens in Madagascar," Mongabay, August 20, 2009, https://news.mongabay.com/2009/08/destruction-worsens-in-madagascar/.
49. Interview with Madagascar National Parks official in Andapa, June 2015 (assuming one log weighs approximately 196 kilograms, as estimated in Hery Randriamalala and Z. Liu, "Bois de rose de Madagascar: Entre démocratie et protection de la nature," *Madagascar Conservation & Development* 5, no. 1 [June 2010]: 10).
50. Interview with Madagascar National Parks official in Andapa, June 2015.
51. Credits are certified through the Climate, Community and Biodiversity Standards (CCB Standards).
52. Cynthia Lalaina Ratsimbazafy, Kazuhiro Harada, and Mitsuru Yamamura, "Forest Resources Use, Attitude, and Perception of Local Residents Towards Community Based Forest Management: Case of the Makira Reducing Emissions from Deforestation and Forest Degradation (REDD) Project, Madagascar," *Journal of Ecology and the Natural Environment* 4, no. 13 (October 2012): 321–332.
53. Although he continued to note, with a slight smile, "*Zegny raha mahagasy atsika*"—"That's what makes us Malagasy!"
54. Ratsimbazafy, Newton, and Ringuet, "Timber Island," 75.
55. Ibid., 30.
56. Ibid., 53.
57. This is likely putting it mildly and not from the perspective of those on the "receiving" end. As Eva Keller notes of her interlocutors around Masoala National Park, "the most sympathetic view of the park was that the development projects offered by the park were utterly insufficient to make up for the losses and that the way forward was to enhance and multiply such projects." The least sympathetic view, in contrast, was the perception of blatant theft of land and livelihood. Keller, *Beyond the Lens of Conservation,* 149.
58. Jean-François Bayart, *The State in Africa: The Politics of the Belly* (Malden, MA: Polity, 2009); Paul Robbins, "The Rotten Institution: Corruption in Natural Resource Management,"

Political Geography 19, no. 4 (May 2000): 423–443; William Reno, *Corruption and State Politics in Sierra Leone* (Cambridge, UK: Cambridge University Press, 1995).

59. And here we see parallels with development economist William Easterly's observation regarding the "elusive quest for growth." William Easterly, *The Elusive Quest for Growth: Economists' Adventures and Misadventures in the Tropics* (Cambridge: MIT Press, 2001).

5. speculating in species

1. Comments made by the Tanzanian delegation on behalf of eleven SADC members at the 18th Conference of the Parties of CITES in Geneva, Switzerland, August 17–28, 2019.
2. "Trade in Elephant, Giraffe and Rhino: 3 African Countries to Take On CITES Rulings," *Africa Geographic,* November 4, 2019; Mmoniemang Motsamai, "Southern Africa: SADC Loses Big at CITES," All Africa, August 27, 2019, https://allafrica.com/stories/201908280702.html.
3. Bram Büscher, "Inverted Commons: Africa's Nature in the Global Imagination," *RCC Perspectives* 5 (2012): 31–38.
4. John Mbaria and Mordecai Ogada, *The Big Conservation Lie: The Untold Story of Wildlife Conservation in Kenya* (Auburn, WA: Lens&Pens, 2016).
5. "*bǎohù xīnzhèng qiǎorán yǐn fà hóngmù jiè qiáng zhèn*" (Protections under New Laws quietly triggered a strong earthquake in the rosewood market), *Sina Collection,* October 11, 2016.
6. Annette Hübschle, "A Game of Horns: Transnational Flows of Rhino Horn" (PhD diss., University of Cologne, 2016); Annette Hübschle, "The Social Economy of Rhino Poaching: Of Economic Freedom Fighters, Professional Hunters and Marginalized Local People," *Current Sociology* 65, no. 3 (2017): 427–447.
7. China's National People's Congress "Decision of the Standing Committee of the National People's Congress on a Complete Ban of Illegal Wildlife Trade and the Elimination of the Unhealthy Habit of Indiscriminate Wild Animal Meat Consumption for the Protection of Human Life and Health," February 24, 2020. http://www.npc.gov.cn/englishnpc/lawsoftheprc/202003/e31e4fac9a9b4df693d0e2340d016dcd.shtml.
8. United Nations Office on Drugs and Crime, *World Wildlife Crime Report* (New York: United Nations, 2016), 46.
9. Ibid.
10. Monique Sosnowski, "Black Markets: A Comparison of the Illegal Ivory and Narcotic Trades," *Deviant Behavior* 41, no. 4 (2020): 8, citing Michael 't Sas-Rolfes, Brendan Moyle, and Daniel Stiles, "The Complex Policy Issue of Elephant Ivory Stockpile Management," *Pachyderm* 55 (January–June 2014): 62–77.
11. Hong Kong Government, Ivory trade, Annex 2, http://www.info.gov.hk/gia/general/201406/04/P201406040449.htm, cited in 't Sas-Rolfes, Moyle, and Stiles, "Complex Policy Issue."
12. Duan Biggs et al., "Legal Trade of Africa's Rhino Horns," *Science* 339 (2013): 1038.
13. Ibid.
14. Daniel W. S. Challender, Stuart R. Harrop, and Douglas C. MacMillan, "Understanding Markets to Conserve Trade-Threatened Species in CITES," *Biological Conservation* 187 (July 2015): 249–259.
15. "*bànnián biāo zhǎng 50%! Zhège bǐ chǎo fáng gèng xiōngměng*" (Fifty percent in the first half of the year! This is more ferocious than real estate!), *China Central Television,* 2017.

16. Sosnowski, "Black Markets," 10.
17. Monica Medina, "The White Gold of Jihad," *New York Times,* September 30, 2013.
18. Proposal submitted by Eswatini at the 18th Conference of the Parties of CITES in Geneva, Switzerland, August 17–28, 2019, https://cites.org/sites/default/files/eng/cop/18/prop/060319/E-CoP18-Prop-08.pdf.
19. Hübschle, "Game of Horns," 50.
20. Brian Handwerk, "Elephants Attack as Humans Turn Up the Pressure," *National Geographic,* June 3, 2005.
21. Linda Givetash, "It's Not Poachers Killing Elephants in Botswana. That Worries Conservationists," NBC News, July 20, 2020.
22. Rosaleen Duffy et al., "Why We Must Question the Militarisation of Conservation," *Biological Conservation* 232 (April 2019): 66–73. See also Francis Massé, Elizabeth Lunstrum, and Devin Holterman, "Linking Green Militarization and Critical Military Studies," *Critical Military Studies* 4, no. 2 (2018): 201–221, and Bram Büscher and Maano Ramutsindela, "Green Violence: Rhino Poaching and the War to Save Southern Africa's Peace Parks," *African Affairs* 115, no. 458 (January 2016): 1–22.
23. Elizabeth Lunstrum, "Green Militarization: Anti-poaching Efforts and the Spatial Contours of Kruger National Park," *Annals of the Association of American Geographers* 104, no. 4 (2014): 816–832, 822.
24. Wendy Annecke and Mmoto Masubelele, "A Review of the Impact of Militarisation: The Case of Rhino Poaching in Kruger National Park, South Africa," *Conservation & Society* 14, no. 3 (January 2016): 195–204.
25. Xuehong Zhou et al., "Elephant Poaching and the Ivory Trade: The Impact of Demand Reduction and Enforcement Efforts by China from 2005–2017," *Global Ecology and Conservation* 16 (October 2018): e00486.
26. Timothy C. Haas and Sam M. Ferreira, "Combating Rhino Horn Trafficking: The Need to Disrupt Criminal Networks," *PLoS ONE* 11, no. 11 (2016): e0167040.
27. "Rosy Times Ahead for Rosewood Furniture," *China Daily,* January 17, 2011.
28. This study was conducted using data from Ghana, a range country that imposed felling and export bans backed by a CITES Appendix III listing in 2016, up-listed to Appendix II in 2017. William Kwadwo Dumenu, "Assessing the Impact of Felling / Export Ban and CITES Designation on Exploitation of African Rosewood (*Pterocarpus erinaceus*)," *Biological Conservation* 236 (August 2019): 124–133.
29. Those I interviewed estimated that sales had been reduced by 30–50 percent because of this massive political shift. One manufacturer indicated that Xi's campaign was the primary reason why rosewood imports and sales began to drop after 2014, following the CITES-related price spikes the year earlier.
30. Ross Harvey, Chris Alden, and Yu-Shan Wu, "Speculating a Fire Sale: Options for Chinese Authorities in Implementing a Domestic Ivory Trade Ban," *Ecological Economics* 141 (November 2017): 22–31.
31. United Nations Office on Drugs and Crime, *World Wildlife Crime Report* (New York: United Nations, 2020).
32. "*zhōngguó yǎng hǔ dì yī rén*" (The first person to raise tigers in China), *Sohu News,* May 5, 2014, http://news.sohu.com/20140505/n399158713.shtml.
33. Kristin Nowell, "Tiger Farms and Pharmacies: The Central Importance of China's Trade Policy for Tiger Conservation," in *Tigers of the World: The Science, Politics, and Conservation of* Pan-

thera tigris, ed. Ronald Tilson and Philip J. Nyhus (Norwich, NY: William Andrew, 2010), 463–475.

34. 't Sas-Rolfes, Moyle, and Stiles, "Complex Policy Issue."
35. Rene Ebersole, "Florida Opened a Fake Alligator Farm to Catch Poachers," *National Geographic,* November 15, 2017.
36. Zhihua Zhou and Zhigang Jiang, "International Trade Status and Crisis for Snake Species in China," *Conservation Biology* 18, no. 5 (October 2004): 1386–1394; Vincent Nijman and Chris R. Shepherd, "The Role of Thailand in the International Trade in CITES-Listed Live Reptiles and Amphibians," *PloS ONE* 6, no. 3 (2011): e17825; Jessica A. Lyons and Daniel J. D. Natusch, "Wildlife Laundering through Breeding Farms: Illegal Harvest, Population Declines and a Means of Regulating the Trade of Green Pythons (*Morelia viridis*) from Indonesia," *Biological Conservation* 144, no. 12 (December 2011): 3073–3081; Fabio Mattioli, Claudia Gili, and Franco Andreone, "Economics of Captive Breeding Applied to the Conservation of Selected Amphibian and Reptile Species from Madagascar," *Natura—Società italiana di Scienze naturali e Museo civico di Storia Naturale di Milano* 95, no. 2 (2006): 67–80.
37. Laura Tensen, "Under What Circumstances Can Wildlife Farming Benefit Species Conservation?," *Global Ecology and Conservation* 6 (April 2016): 286–298; Mattioli, Gili, and Andreone, "Economics of Captive Breeding."
38. A notable exception is India.
39. "A Legal Trade in Rhino Horn," Save the Rhino, December 21, 2018, https://www.savetherhino.org/thorny-issues/legal-trade-in-rhino-horn/.
40. See, for example, Judy A. Mills, *Blood of the Tiger: A Story of Conspiracy, Greed, and the Battle to Save a Magnificent Species* (Boston: Beacon Press, 2015), chapter 8. For another example, see Chris Alden and Ross Harvey, "South African Proposal to Breed Wildlife for Slaughter Courts Disaster," *The Conversation,* June 14, 2020, which observes that wildlife farming "reduces wild animals to mere consumables."
41. Chris Coggins, *The Tiger and the Pangolin: Nature, Culture, and Conservation in China* (Honolulu: University of Hawai'i Press, 2003), 53–59.
42. Ibid., 67.
43. Coggins clarifies: "Though medieval Europeans also believed that particular parts of certain wild plants and animals held medicinal value, this belief did not develop into the complex, systematic ethnoscience that Chinese medicine has become." Ibid., 69. Also, quoting Esther Cohen, he notes that in medieval European uses, animals were "not only inferior to the human in a hierarch of government, [they were] also further removed from divinity. . . . The use of animal symbolism had nothing theological about it." Esther Cohen, "Animals in Medieval Perceptions: The Image of the Ubiquitous Other," in *Animals and Human Society: Changing Perspectives,* ed. Aubrey Manning and James Serpell (New York: Routledge, 1994), 61.
44. Yu Xiao, "China's Ivory Market after the Ivory Trade Ban in 2018," TRAFFIC Briefing Paper (Cambridge, UK: TRAFFIC, September 2018); Wander Meijer et al., "Demand under the Ban: China Ivory Consumption Research Post-Ban 2018" (Beijing: TRAFFIC and World Wildlife Fund, 2018); Hsun-Wen Chou, "China's Ivory Auction Market: A Comprehensive Analysis of Legislation, Historical Data and Market Survey Results," TRAFFIC Report (Cambridge, UK: TRAFFIC, September 2018).
45. Xiangying Shi et al., "Public Perception of Wildlife Consumption and Trade during the COVID-19 Outbreak," *Biodiversity Science* 28, no. 5 (2020): 630–643.

6. pluralizing environmentalism

1. Ving Wu, "Ant Forest Plants Trees for Sustainable Behavior," *Revolve,* March 20, 2020.
2. Meng Zhang, "Financing Market-Oriented Reforestation: Securitization of Timberlands and Shareholding Practices in Southwest China, 1750–1900," *Late Imperial China* 38, no. 2 (2017): 109–151; Joseph P. McDermott, *The Making of a New Rural Order in South China,* vol. 1, *Village, Land, and Lineage in Huizhou, 900–1600* (Cambridge, UK: Cambridge University Press, 2013); Ian Miller, "Roots and Branches: Woodland Institutions in South China, 800–1600" (PhD diss., Harvard University, 2015).
3. Stanley Dennis Richardson, *Forests and Forestry in China: Changing Patterns of Resource Development* (Washington, DC: Island Press, 1990).
4. Jonathan Watts, *When a Billion Chinese Jump: Voices from the Frontline of Climate Change* (London: Faber and Faber, 2010); Chi Chen et al., "China and India Lead in Greening of the World through Land-Use Management," *Nature Sustainability* 2, no. 2 (2019): 122–129.
5. Chen et al., "China and India Lead in Greening of the World."
6. Throughout this chapter, I use the term "reforestation" to cover both reforestation and afforestation efforts. This is for brevity and also because, given China's long history of human-environment interactions, it is very difficult to say which parts of China were "naturally" forested (and thus *re*forested) and which were never forested (and thus *af*forested).
7. Damian Carrington, "Tree Planting 'Has Mind-Blowing Potential' to Tackle Climate Crisis," *Guardian,* July 4, 2019, citing Jean-Francois Bastin et al., "The Global Tree Restoration Potential," *Science* 365, no. 6448 (2019): 76–79.
8. Simon L. Lewis et al., "Comment on 'The Global Tree Restoration Potential,'" *Science* 366, no. 6463 (2019); Yiwen Zeng et al., "Economic and Social Constraints on Reforestation for Climate Mitigation in Southeast Asia," *Nature Climate Change* 10, no. 9 (2020): 842–844; M. E. Fagan, "A Lesson Unlearned? Underestimating Tree Cover in Drylands Biases Global Restoration Maps," *Global Change Biology* 26, no. 9 (September 2020): 4679–4690; Xiaowei Tong et al., "Increased Vegetation Growth and Carbon Stock in China Karst via Ecological Engineering," *Nature Sustainability* 1, no. 1 (2018): 44–50; Jianxiao Zhu et al., "Increasing Soil Carbon Stocks in Eight Permanent Forest Plots in China," *Biogeosciences* 17, no. 3 (2020): 715–726; Xiaowei Tong et al., "Forest Management in Southern China Generates Short Term Extensive Carbon Sequestration," *Nature Communications* 11, no. 1 (2020): 1–10.
9. Michael Holtz, "China Spent US$100 Billion on Reforestation. So Why Does It Have 'Green Deserts'?," *Christian Science Monitor,* June 18, 2017; Yifei Li and Judith Shapiro, *China Goes Green: Coercive Environmentalism for a Troubled Planet* (Cambridge, UK: Polity Press, 2020).
10. Ecological civilization (*shēngtài wénmíng*), as discussed further in a subsequent note, is the broad rhetorical and policy umbrella under which all of China's state-sponsored environmental initiatives are pursued. Simultaneously political, cultural, and highly technical, the concept has been referred to as a "sociotechnical imaginary" invoking China's multimillennia civilizational heritage as part of the solution to the planet's environmental future. Mette Halskov Hansen, Hongtao Li, and Rune Svarverud, "Ecological Civilization: Interpreting the Chinese Past, Projecting the Global Future," *Global Environmental Change* 53 (November 2018): 195–203.
11. Robert Weller, *Discovering Nature: Globalization and Environmental Culture in China and Taiwan* (Cambridge, UK: Cambridge University Press, 2006); Robert B. Marks, *China: An Environmental History* (Lanham, MD: Rowman & Littlefield, 2017).

12. You-tien Hsing, *The Great Urban Transformation: Politics of Land and Property in China* (Oxford: Oxford University Press, 2010); Xuefei Ren, *Urban China* (Hoboken, NJ: John Wiley & Sons, 2013); Li Zhang, Richard LeGates, and Min Zhao, *Understanding China's Urbanization: The Great Demographic, Spatial, Economic, and Social Transformation* (Cheltenham, UK: Edward Elgar, 2016).
13. Judith Shapiro, *Mao's War against Nature: Politics and the Environment in Revolutionary China* (Cambridge, UK: Cambridge University Press, 2001).
14. Richardson, *Forests and Forestry in China,* 18.
15. Ibid. This occurred alongside massive deforestation during the Great Leap Forward (1958–1960), at which time forests were cut to fuel backyard furnaces for highly inefficient industrial production, and also during the Cultural Revolution (1966–1976), at which time forests were cleared to make way for increased grain production. Marks, *China: An Environmental History.*
16. Daniel Rechtschaffen, "How China's Growing Deserts Are Choking the Country," *Forbes,* September 18, 2017.
17. As with Mao, Deng's planting occurred alongside massive deforestation, which was an unintended consequence of dismantling collectivized agriculture and additional building campaigns. Marks, *China: An Environmental History.*
18. *Dèngxiǎopíng de zhíshù qíngjié* ("Deng Xiaoping's Complex of Planting Trees"), *Guang'an Daily,* March 10, 2017.
19. Ibid.
20. Laura Oliver, "China Has Sent 60,000 Soldiers to Plant Trees," World Economic Forum, February 16, 2018; "China to Create New Forests Covering Area Size of Ireland," *China Daily,* January 4, 2018.
21. Jiang Yifan, "14th Five Year Plan: China's Carbon-Centered Environmental Blueprint," *China Dialogue,* March 25, 2021.
22. "China announces massive greening plan to achieve carbon goals," *Xinhua,* August 24, 2021. http://www.news.cn/english/2021-08/24/c_1310146397.htm.
23. Tree mortality, stunting, and lowering of the water table resoundingly point to the project's shortcomings. More than this, the trees being planted—fast-growing varieties that can be harvested for wood pulp—do not contribute to restoring local biodiversity and often strain scarce water resources. "Such plantations," a prominent environmental historian affirms, "cannot be considered 'forests' in the sense of preserving biodiversity." Marks, *China: An Environmental History,* 340. Echoing this viewpoint, others have labeled China's tree planting as a "mask," "an expensive band aid," or blatant "propaganda" glossing over the deeper impacts to biodiversity that China's development is inflicting. Evan Ratliff, "The Green Wall of China," *Wired,* April 1, 2003.
24. In contrast to the criticism discussed in the previous note, China's State Forestry Administration declares the project an overwhelming success. Acknowledging initial shortcomings, the administration nonetheless reports that the project has reduced sandstorms by 20 percent and desertification by nearly five thousand miles in recent years. It is projected that much of the arid land can be restored to a productive and sustainable state by 2050. In the Beijing region in particular, even though some trees remain stunted or have died, forest cover has reportedly increased from around 3 percent at the middle of the twentieth century to more than 40 percent today. Shen Guofang, "The Harmonious Development of Humans and Nature," in *Chinese Perspectives on the Environment and Sustainable Development,* ed. Ye Wenhu, trans. Christopher Heselton (Leiden: Brill, 2013), 121–165. In a recent report to the United Nations,

Chinese officials optimistically predicted that the effort will "terminate expansion of new desertification caused by human factors" within a decade.

25. Vaclav Smil, *The Bad Earth: Environmental Degradation in China* (London: Routledge, 2015).
26. Richardson, *Forests and Forestry in China,* 191.
27. Marks, *China: An Environmental History,* 340.
28. With the third-highest land coverage of eucalyptus in the world (after Brazil and India), China contains 4.5 million hectares of eucalyptus plantations (more than 6 percent of the country's total planted forests). Yaojian Xie et al., "Advances in Eucalypt Research in China," *Frontiers of Agricultural Science and Engineering* 4, no. 4 (2017): 380.
29. As early as the 1950s, large-scale eucalyptus plantations were established throughout China in an attempt to afforest barren landscapes, with little knowledge of the negative consequences. Since this time, more than three hundred species of eucalyptus have been introduced in the country, and more than two hundred species have been cultivated for afforestation. Y. Xie, "Research Progress on Eucalyptus Breeding and Its Strategy in China," *World Forestry Research* 24, no. 4 (2011): 50–54.
30. The tree can grow three to five centimeters in a day and more than one meter per month. The average rotation is five to eight years, but in some places in China it can be as little as three to four years. Bai Jiayu and Gan Siming, "Eucalyptus Plantations in China," Food and Agriculture Organization of the United Nations, http://www.fao.org/docrep/005/AC772E/ac772e04.htm.
31. The price of rosewood depends primarily on the species and the diameter of the tree when harvested. For example, *Hǎinán huánghuālí* (*Dalbergia odorifera*) that is from a tree sixty-five centimeters in diameter at breast height can be worth more than US$3 million per kilogram (primarily because trees of that species and size are effectively extinct). Roger Arnold, "Corruption, Bloodshed and Death—the Curse of Rosewood," Environmental Investigation Agency, August 16, 2013.
32. These growth estimates are provided by the South India Sandalwood Products Dealers and Exporters Association, as reported in "Aussie Sandalwood Growers Can Cash In Now after 15-Year Wait," *The Straits Times,* February 23, 2017. According to the scientist I interviewed at the Research Institute of Tropical Forestry, Australia is the only place that has the capability of extracting oils from this type of wood.
33. To ensure finicky consumers that they are in fact "free range," these chickens could even be tracked by using a small device, encoding their movements with the same blockchain digital ledger used in cryptocurrency transactions. Xiaowei Wang, *Blockchain Chicken Farm: And Other Stories of Tech in China's Countryside* (New York: Farrar, Straus and Giroux, 2020).
34. Maturation rates are highly dependent on species and their condition of growth. As discussed later in this chapter, experts cultivating *Hǎinán huánghuālí* have reduced the time to harvest to as little as twenty years but more commonly (and for better wood) at least fifty years. Typically, such high-value wood is harvested not all at once but on a rotating basis as individual trees mature.
35. He started his lease twenty years ago for around US$11 per acre over a fifty-year period, with the final leasing price reaching no more than US$21 per acre. Today, a similar lease would cost nearly US$80 per acre—expensive but certainly not a deal breaker for a plantation earning more than US$900 per acre for its tea crop alone.
36. Quoted in Xiaoying Liu, "The Evolution of Civilization and Prospects of an Ecological Culture," in *Chinese Perspectives on the Environment and Sustainable Development,* ed. Ye Wenhu, trans. Christopher Heselton (Leiden: Brill, 2013), 168.

37. Ibid.
38. Ibid., 167.
39. Coraline Goron, "Ecological Civilisation and the Political Limits of a Chinese Concept of Sustainability," *China Perspectives* 4 (2018): 39–52.
40. Kevin Lo, "Review of *Chinese Perspectives on the Environment and Sustainable Development*," *Asian Studies Review* 38 (2014): 4.
41. In 2013, the United Nations Environment Programme adopted a draft decision to promote the concept of ecological civilization in China, marking the recognition and support of the theory and practice of China's ecological civilization in the international community. China's designated hosting of the Conference of the Parties to the Convention on Biological Diversity, scheduled to be held in Kunming in 2020 (but postponed because of the COVID-19 pandemic), provides a more recent example. The concepts of ecological civilization played a strong role in the discourse leading up to the conference.
42. Wenhu, *Chinese Perspectives*, 197.
43. Arthur Hanson, "Ecological Civilization in the People's Republic of China: Values, Action, and Future Needs," Asian Development Bank East Asia Working Paper Series, No. 21 (December 2019), 5. Goron, "Ecological Civilisation," 39.
44. Hanson, "Ecological Civilization."
45. Some have even gone so far as to assert that ecological civilization "cultivates a cultural chauvinism," privileging a Chinese cultural approach over others. Goron, "Ecological Civilisation," 40, citing Gwennaël Gaffric and Jean-Yves Heurtebise, "L'écologie, Confucius et la démocratie," *Écologie & Politique* 2 (2013): 51–61. Yet there is also an emphasis among ecological civilization discourse, especially as it becomes more international, on differentiated approaches based on different cultural considerations. In other words, while political and cultural factors are to be considered in environmental pursuits, they need not be particularly *Chinese* political and cultural factors. Hanson, "Ecological Civilization," 20.
46. See, for example, Martha J. Groom, Gary K. Meffe, and C. Ronald Carroll, *Principles of Conservation Biology* (Sunderland, MA: Sinauer Associates, 2006), chapter 1. Aldo Leopold's evolutionary-ecological land ethic, which emphasizes conservation of the ecological whole, not just specific resources, has been thought to provide some sort of middle ground, but in practice it is most often used to justify the contemporary deep ecology movement, which sits very squarely within the non-anthropocentric camp.
47. Charles C. Mann, *The Wizard and the Prophet: Two Remarkable Scientists and Their Dueling Visions to Shape Tomorrow's World* (New York: Penguin Random House, 2018).
48. William Cronon, *Uncommon Ground: Rethinking the Human Place in Nature* (New York: W. W. Norton, 1995), 60.
49. Cronon, *Uncommon Ground;* Raymond Williams, *The Country and the City* (Oxford: Oxford University Press, 1975); Weller, *Discovering Nature.*
50. Derrida further highlights the persistent fundamental importance of the nature / culture opposition in Western thought through the example of anthropologist Claude Lévi-Strauss, who, according to Derrida, "simultaneously has experienced the necessity of utilizing this opposition and the impossibility of accepting it" (357). Pointing to Lévi-Strauss's seminal work, *The Elementary Structures of Kinship,* Derrida observes that Lévi-Strauss begins this book with the scandalous observation that the incest prohibition is both "natural" (in that it is universal, shared by all human groups in some form or another) and "cultural" (in that it clearly abides by a system of norms and interdicts). Lévi-Strauss is so fascinated by the impossibility of straddling these borders that he calls it a "scandal," yet Derrida reminds

readers that this transgression only appears as such to faithful adherents of nature / culture opposition. "Obviously," Derrida observes, "there is no scandal except within a system of concepts which accredits the difference between nature and culture" (358). Jacques Derrida, *Writing and Difference,* trans. Alan Bass (London: Routledge, 1978).

51. Weller, *Discovering Nature.*
52. "Nature needs half" is a preservationist movement defined by the following call to action: "You know what needs to be done: save nature, end the biodiversity crisis, help protect humanity. What you might not know is how we're going to do it. The best contemporary science and traditional wisdom tell us that *nature needs half.* That may seem like a lot, but we have a plan for how to get there and transform the way society thinks about and benefits from nature." See https://natureneedshalf.org/.
53. J. Donald Hughes, *An Environmental History of the World: Humankind's Changing Role in the Community of Life* (Abingdon, Oxon, UK: Routledge, 2009), 70.
54. Indeed, most languages in the world do not have a word for "nature" in the Western sense, as referring to the nonhuman world. Josep-Maria Mallarach et al., "Implications of the Diversity of Concepts and Values of Nature in the Management and Governance of Protected and Conserved Areas," in *Cultural and Spiritual Significance of Nature in Protected Areas: Governance, Management and Policy,* ed. Bas Verschuuren and Steve Brown (Abingdon, Oxon, UK: Routledge, 2018), 21.
55. Weller, *Discovering Nature,* 25.
56. "Correspondence" or "similitude" refers to the pre-Enlightenment assumption that there is an inherent connection between the world and discourse used to describe it, as discussed by philosopher Michel Foucault in his seminal book *The Order of Things.* In a world characterized by similitude, words (among other signs) are not deemed arbitrary but are reflective of the essence of the things they are enlisted to define. "In its original form . . . language was an absolutely certain and transparent sign for things, because it resembled them," Foucault asserts. "The names of things were lodged in the things they designated, just as strength is written in the body of the lion, regality in the eye of the eagle, just as the influence of the planets is marked upon the brows of men: by the form of similitude." Here, we see an understanding of the world (or "episteme," as Foucault phrases it) not characterized by the nature / culture divide. Michel Foucault, *The Order of Things: An Archaeology of the Human Sciences* (London: Routledge, 2005), 40.
57. Liu, "Evolution of Civilization," 190.
58. Ibid., 172.
59. Ibid., 178.
60. "Bomenplant werkt alleen met divers bos," *NRC,* April 9, 2019.
61. "President Xi Plants Trees, Urges Forestry Development," *China Daily,* April 5, 2016.
62. "Poplar Trees Get New Status under BRI," *China Daily,* September 14, 2018.
63. Unai Pascual et al., "Biodiversity and the Challenge of Pluralism," *Nature Sustainability* 4 (2021): 1–6.

conclusion

1. Malagasy rosewood is from the genus *Dalbergia,* whereas *zǐtán* is from the genus *Pterocarpus.* Of course, the category "genus" itself and the division of life accordingly are largely arbitrary. In fact, because of the discovered similarity between these two genuses, the scientific

community has created a tribe (the taxonomic category above genus and below family) named Dalbergieae to unite the *Dalbergia* and *Pterocarpus* genuses.

2. Hany Besada, Yang Wang, and John Whalley, "China's Growing Economic Activity in Africa," in *China's Integration into the World Economy,* ed. John Whalley (Hackensack, NJ: World Scientific, 2011), 230; Miria Pigato and Wenxia Tang, "China and Africa: Expanding Economic Ties in an Evolving Global Context," World Bank Group, 2015, https://openknowledge.worldbank.org/handle/10986/21788.
3. Shirley Ze Yu, "Why Substantial Chinese FDI Is Flowing into Africa," *London School of Economics Blogs,* April 2, 2021.
4. Harry G. Broadman, *Africa's Silk Road: China and India's New Economic Frontier* (Washington, DC: World Bank, 2006).
5. Pigato and Tang, "China and Africa: Expanding Economic Ties."
6. Reduced trade in these minerals was a result of Section 1502 of the US Dodd-Frank Act, passed in 2010, which contains a "conflict minerals" provision requiring US companies to perform due diligence in order to ensure their imports of tin, tungsten, tantalum, and gold coming from the Congo region are not funding armed groups or human rights abuses.
7. Philippe Tunamsifu Shirambere, "The Democratic Republic of the Congo–China's Deals on Construction of Roads in Exchange of Mines," *Afrika Focus* 33, no. 2 (December 20, 2020).
8. Olayiwola Abegunrin and Charity Manyeruke, "China-Zimbabwe Relations: A Strategic Partnership?," in *China's Power in Africa: A New Global Order* (Cham, Switzerland: Palgrave Macmillan, 2020), 95–113.
9. Broadman, *Africa's Silk Road,* 73.
10. Howard W. French, *China's Second Continent: How a Million Migrants Are Building a New Empire in Africa* (New York: Vintage Books, 2014).
11. As of January 2021, 140 countries had officially signed a memorandum of understanding with China. Christoph Nedopil, "Countries of the Belt and Road Initiative (BRI)" (Shanghai: FISF Fudan University, Green Finance & Development Center, 2021), https://green-bri.org/countries-of-the-belt-and-road-initiative-bri/.
12. Respectively cited in Keith Bradsher, "At Davos, the Real Star May Have Been China, Not Trump," *New York Times,* January 28, 2018; Jessica Meyers, "China's Belt and Road Forum Lays Groundwork for a New Global Order," *Los Angeles Times,* May 15, 2017; James A. Millward, "Is China a Colonial Power?," *New York Times,* May 4, 2018; Jane Perlez and Yufan Huang, "Behind China's $1 Trillion Plan to Shake Up the Economic Order," *New York Times,* May 13, 2017.
13. Wang Gungwu, *The Chinese Overseas: From Earthbound China to the Quest for Autonomy* (Cambridge, MA: Harvard University Press, 2009).
14. Lynn Pan, ed., *The Encyclopedia of the Chinese Overseas* (Cambridge, MA: Harvard University Press, 1999).
15. "China Joins the WTO—at Last," BBC News, December 11, 2001.
16. Ai Weiwei, "Think 'Sanctions' Will Trouble China? Then You're Stuck in the Politics of the Past," *Guardian,* August 6, 2020.
17. George H. W. Bush, remarks at the Yale University commencement ceremony in New Haven, Connecticut, May 27, 1991.
18. Michael Schuman, "The Undoing of China's Economic Miracle," *Atlantic,* January 10, 2021.
19. Ronald Coase and Ning Wang, *How China Became Capitalist* (Basingstoke, UK: Springer, 2012).

20. Ai, "Think 'Sanctions' Will Trouble China?"
21. Ching Kwan Lee, *The Specter of Global China: Politics, Labor, and Foreign Investment in Africa* (Chicago: University of Chicago Press, 2017).
22. Ibid., 28–29.
23. Ibid., 155.
24. Maria Adele Carrai, *Sovereignty in China: A Genealogy of a Concept since 1840* (Cambridge, UK: Cambridge University Press, 2019).
25. Building on this sentiment, Professor Zhao Lei describes the Belt and Road Initiative as the start of a new Chinese-style discourse on international relations that emphasizes inclusiveness and tolerance and embodies an "anti-polarizing" (*fēi jí huà*) tendency in international relations. China's "civilizational values," Zhao continues, are built on respect for "the diversity of the world's civilizations and their unique national development models." Samuli Seppänen, "Performative Uses of Sovereignty in the Belt and Road Initiative," in *International Governance and the Rule of Law in China under the Belt and Road Initiative,* ed. Yun Zhao (Cambridge, UK: Cambridge University Press, 2018), 52.
26. Seppänen, "Performative Uses of Sovereignty in the Belt and Road Initiative"; Kenneth Kalu, "'Respect' and 'Agency' as Driving Forces for China–Africa Relations," *Place Branding and Public Diplomacy* (2020): 1–12.
27. Gyude Moore, "China in Africa: An African Perspective," talk given at the University of Chicago Paulson Institute's Contemporary China Speaker Series, March 5, 2019.
28. "Clinton Warns against 'New Colonialism' in Africa," Reuters, June 11, 2011; Deborah Bräutigam, "Is China the World's Loan Shark?" *New York Times,* April 26, 2019.
29. Adva Saldinger, "African Leaders Question US Position on China at Investment Event," *Devex News,* October 19, 2020.
30. Fernando Ascensão et al., "Environmental Challenges for the Belt and Road Initiative," *Nature Sustainability* 1, no. 5 (2018): 206–209.
31. Andre Gunder Frank, *ReOrient: Global Economy in the Asian Age* (Berkeley: University of California Press, 1998).
32. Lee, *Specter of Global China.*
33. Coase and Wang, *How China Became Capitalist,* 62.
34. For the full text of the keynote speech of the Belt and Road Forum for International Cooperation, see http://www.xinhuanet.com/english/2017-05/14/c_136282982.htm.
35. Cynthia Ratsimbazafy, D. Newton, and S. Ringuet, "Timber Island: The Rosewood and Ebony Trade of Madagascar," TRAFFIC Report, November 2016.
36. On top of this, Liang was also commissioned to add a novelty Western-styled mansion within one of Emperor Qianlong's many rosewood palaces.
37. Interestingly, the poem inscription on the painting (written by Emperor Qianlong himself) indicates that the creature depicted is incorrectly thought to be from Cochin (contemporary Vietnam) rather than Madagascar. But where exactly this creature is from is of little import to the emperor. The lemur need not signify anything more than the very generalizable "global exotic" in order for this piece of artwork to symbolize a manifesto of global China. The inscription on the painting is roughly translated as follows:

> A lemur is born and lives in Vietnam and call himself "Guaran."
> He is still enjoying a cheerful life, but it is difficult to return to his normal life with the group of lemurs.

He used to live with the group of lemurs, who are kind to each other and help each other with respect to the order of seniority.
The willow grove is different now than it used to be, and the prince of the palace will tease him.
The king is a virtuous man in a tall building.
If not because his beautiful fur is suitable for making a blanket,
why should he die in the woods for no reason?

(*Translation provided by Banglong Zhu.*)

As demonstrated by the text, the emperor both grapples with the motivation for lemur slaughter and imposes Confucian ideals of harmony on their social lives.

38. Aihwa Ong, "What Marco Polo Forgot: Contemporary Chinese Art Reconfigures the Global," *Current Anthropology* 53, no. 4 (2012): 473. She further concludes that "by assembling and juxtaposing disparate elements (West-East, past and present, culture and technology, etc.) in global spaces of encounter, modern Chinese art is anticipatory of a new global, one that embraces inevitable heterogeneity, subversion, and uncertainty." Ibid., 475.
39. Rivonala Razafison and Malavika Vyawahare, "Madagascar Minister Calls Protected Areas a 'Failure,' Seeks People-Centric Approach," Mongabay, August 20, 2020.
40. Bram Büscher, "Inverted Commons: Africa's Nature in the Global Imagination," *RCC Perspectives* 5 (2012): 31–38.
41. For more on convivial futures, see Bram Büscher and Robert Fletcher, *The Conservation Revolution: Radical Ideas for Saving Nature beyond the Anthropocene* (New York: Verso, 2020).
42. Donna J. Haraway, *Staying with the Trouble: Making Kin in the Chthulucene* (Durham, NC: Duke University Press, 2016), 2; Anna Lowenhaupt Tsing, *The Mushroom at the End of the World: On the Possibility of Life in Capitalist Ruins* (Princeton, NJ: Princeton University Press, 2015), 3.
43. "Madagascar: 60 Million Trees to Be Planted for 60 Years of Independence," *North Africa Post,* March 16, 2020.
44. Ibid.

acknowledgments

This book reconsiders endangered species conservation in light of the perspectives of the people involved in the world's largest illicit wildlife trade: rosewood. I am so grateful to those whose lives have been impacted by this trade for sharing their knowledge and experiences with me.

In Madagascar, I would like to thank so many people whose stories appear in the book, but I will refrain from naming certain names, given the political sensitivity of the topic. I owe special thanks to two friends-turned-research-assistants who took me around the villages and tolerated my infinite questions about both the very exciting and the very mundane details of the trade. This research simply could not have been completed if not for them. I am also greatly indebted to my host family—Nany and Dada, Fanatenana and Madrindra—in Montasoa, and my closest Malagasy neighbors from my Peace Corps days, especially Mahitso, Lydia, and Arnold, who taught me the language and culture. I would also like to thank those with whom I worked as a Peace Corps volunteer at the nongovernmental organization Macolline, including Marie Hélène Kam Hyo Zschocke, Marie Mélanie Voahangiarisoa, Belucien, and Ericlin. Their work with reforestation is a model for the region.

In China, I am first and foremost grateful to my father-in-law, Zhu Banglong, who served as my translator and native Shanghai informant, taking me around to various rosewood dealers. My uncle-in-law, Deng Ronghua, luckily also occasionally attended these field trips, supplementing my own list of questions with many more that I would have never thought to ask. My mother-in-law, Deng Meihua, was so helpful both in China and at home; neither my PhD nor my fieldwork in China would have been possible if not for her. I am also deeply indebted to Xu Daping and Hong Zhou from the

Chinese Academy of Sciences, who took my husband and me around to visit various rosewood plantations. Their work on rosewood reforestation is contributing in no small measure to saving the species. I am also grateful to Michael Lu and his family in Guangzhou, who hosted my husband and me and put us in contact with rosewood furniture manufacturers in the region. Thanks go as well to my collaborators in China, including Liu Xuehua, Li Zhouyuan, Jiao Xiaoqiang, Li Xiaodan, Guo Xiaoxia, Wang Chong, Zhang Fusuo, Guo Xiaona, and Chen Ruishan.

At the University of California, Berkeley, I am especially grateful to Nancy Peluso, who brought me to the PhD program and taught me all I know about political ecology. Her support and vision kept me going through the years, and our land lab meetings were one of the highlights of my doctoral work. I am also grateful to Aihwa Ong for opening up an entirely new way of thinking about globalization and the non-Western world, a legacy on which I hope to build. Kate O'Neill was an early mentor in my PhD program; her work has been such an inspiration to me and will continue to be, moving forward. Nathan Sayre, Michael Watts, Justin Brashares, Isha Ray, and Claire Kremen also helped guide me to where I am now. Finally, I am indebted to the Center for African Studies and the Center for Chinese Studies for providing research and writing support, as well as Brian Klein, Matt Libassi, Juliet Lu, Jane Flegal, Lisa Kelly, Ashton Wesner, Laura Dev, Tracy Hruska, Alice Kelly, Mez Baker-Médard, and other land lab members for comradery and constructive comments on my drafts.

At my current home in the Environmental Policy Group of Wageningen University, I am so lucky to benefit from the inspiring leadership of Simon Bush and the ongoing legacy pioneered by Gert Spaargaren, Peter Oosterveer, Arthur Mol, Aarti Gupta, and Bas van Vliet. Kris van Koppen has been especially helpful in cultivating ongoing discussions on nature conservation, and I have come to think of Ingrid Boas as a gracious mentor. I am also deeply thankful for the great discussions shared with my new colleagues Machiel Lamers, Sigrid Wertheim-Heck, Judith van Leeuwen, Sanneke Kloppenburg, Mattijs Smits, Mary Greene, Annet Pauwelussen, Hilde Toonen, Maartje van der Knaap, Nowella Anyango-van Zwieten, Jillian Student, Hanne Wiegel, Paulina Rosero Anazco, Gertjan Hofstede, Erika Speelman, Jessica Duncan, Maira Jong van Lier, Zhu Xueqin, Mindy Schneider, Robert Fletcher, and Bram Büscher. Jin Qian was especially helpful in reviewing the manuscript and providing further insights and research support.

At Harvard University Press, I am extremely grateful to my editor, Joseph Pomp, for taking the manuscript so far beyond where it began. I would also like to thank Mihaela Pacurar and Eric Mulder for helping pull everything together seamlessly and making the imagery come alive and Pat Harris for expertly polishing the text. Finally, I am indebted to two anonymous reviewers who provided expert suggestions for improving the manuscript. The introduction and Chapter 3 build on ideas first discussed in "Rosewood Occidentalism and Orientalism in Madagascar," *Geoforum* 86 (November 2017): 1–12. The introduction and Chapter 1 also touch on material first presented in "China's Rosewood Boom: A Cultural Fix to Capital Overaccumulation," *Annals of the American Association of Geographers* 110, no. 1 (2020), 277–296. Portions of Chapter 2 were first published as "Hot Money, Cold Beer: Navigating the Vanilla and Rosewood Export Economies in Northeastern Madagascar," *American Ethnologist* 45, no. 2 (May 2018): 253–267.

On a more personal note, I am eternally indebted to my mother and father, Carla and Greg, for being there for me, in their very different ways. I am above all grateful for my mother's ability to see the world in a way that no other can and for my father's poetic instincts. My older sister and brother, Tamzyn and Oliver, are here too in this book, very close to who I am and who I am becoming no matter how far apart we are. Then there are my two boys, Theophilus Pacific and Malakai Atlantic, who I must thank for taking me away from my work and reminding me of the few things one might call "real" in this world. And my last and greatest gratitude goes to my husband, George Zhu, whose ideas have formed the basis of this book as much as my own. He has read through countless drafts, has contributed countless rhetorical flourishes and critical insights, and has continuously challenged me to challenge my readers to think beyond the obvious in making sense of the world. How boring this book, this life, would be if it were not for you.

index

Note: Italicized page numbers indicate illustrative material.